# Biomembrane Protocols
## I. Isolation and Analysis

T0137765

# Methods in Molecular Biology

*John M. Walker,* Series Editor

**Methods in Molecular Biology • 19**

# Biomembrane Protocols

## *I. Isolation and Analysis*

Edited by

## *John M. Graham*

*Merseyside Innovation Centre, Liverpool, UK*

## *Joan A. Higgins*

*Department of Molecular Biology and Biotechnology,
University of Sheffield, UK*

**Humana Press** ✳ **Totowa, New Jersey**

© 1993 Humana Press Inc.
999 Riverview Drive, Suite 208
Totowa, New Jersey 07512

Printed in the United States of America. 9 8 7 6 5 4 3 2 1

Library of Congress Cataloging in Publication Data
Main entry under title:

Methods in molecular biology.

Biomembrane protocols / edited by John M. Graham, Joan A. Higgins.
        p.   cm. — (Methods in molecular biology ; 19–)
    Includes index.
    Contents: 1. Isolation and analysis.
    ISBN 0-89603-236-1
        1. Membranes (Biology) I. Graham, John M. II. Higgins, Joan A.
    III. Series: Methods in molecular biology (Totowa, NJ) ; 19–
    QH601.B5257  1993
    574.87'5—dc20                                    93-13471
                                                        CIP

# Contents

*vii*

# Preface

There have been many important advances in our understanding of biological membrane structure and function over the last decade. Much of this progress has been driven by the development of new techniques for studying membrane components and their interactions. Traditionally, the investigation of membranes has largely occurred within the domains of biochemistry, physical chemistry, and cell biology; but many of the most significant advancements have resulted from an expansion into other disciplines, such as molecular biology, immunology, and the clinical sciences. In these two volumes we have attempted to bring together some of these techniques—to combine the old and the new. *Biomembrane Protocols: I. Isolation and Analysis* is concerned exclusively with the isolation and compositional analysis of membranes, whereas the companion volume, *Biomembrane Protocols: II. Architecture and Function,* deals with their architecture and activities.

The aim of each chapter is to provide detailed technical and methodological information to allow the reader to perform the technique successfully, without the need to consult other texts. Detailed reviews of membrane structure and function are not included, except when they are relevant to the choice or efficacy of a particular procedure. In addition to the Materials and Methods sections, each chapter has a Notes section that explains the reasons for taking certain steps and provides practical tips for applying the techniques to other systems. Of necessity, some chapters may diverge from this simple format; some deal with techniques that require adapting to suit a specific application, whereas others include a combination of different techniques. In these cases, the text may be more discursive; nevertheless detailed examples are always given. Sometimes the Notes section is considerably expanded to provide worked-out examples or to help in interpreting experimental results.

The first half of this volume is concerned with the isolation of subcellular fractions, including nuclei, nuclear envelopes, mitochondria, lysosomes, peroxisomes, Golgi membranes, plasma membranes, coated vesicles, and the components of the endocytic compartment from animal tissues and cells. The isolation of membranes from plant tissue and bacteria are also covered. The main technique used in these chapters on isolation is centrifugation, but the use of antibodies and electrophoresis are included. The second half of the volume is concerned with the compositional analysis of membranes. Topics include the extraction and quantitation of lipids; the separation and analysis of proteins; and the isolation and characterization of glycosylated molecules, glycoproteins, proteoglycans, and glycolipids.

The study of biological membranes covers an enormous range of topics and employs a great diversity of techniques. We have attempted to cover most of the techniques that are likely to be of substantial interest to most biochemists and cell biologists. Inevitably an important methodology or two will have escaped treatment; the editors will be grateful to receive comments and suggestions from our readers that might be incorporated in future editions.

*John M. Graham*
*Joan A. Higgins*

# Contributors

HAFEZ A. AHMED • *Department of Biochemical Medicine, St. George's Hospital Medical School, London, UK*

ELAINE M. BAILYES • *Department of Clinical Biochemistry, University of Cambridge, Addenbrooke's Hospital, Cambridge, UK*

SAMANTHA J. BRADD • *Department of Cardiothoracic Surgery, National Heart and Lung Institute, Heart Science Centre, Harefield, UK*

IAN J. CARTWRIGHT • *Department of Molecular Biology and Biotechnology, The University of Sheffield, South Yorkshire, UK*

JOE CORBETT • *Department of Cardiothoracic Surgery, National Heart and Lung Institute, Heart Science Centre, Harefield, UK*

ANTHONY P. CORFIELD • *University Department of Medicine Laboratories, Bristol Royal Infirmary, Bristol, UK*

MICHAEL J. DUNN • *Department of Cardiothoracic Surgery, National Heart and Lung Institute, Heart Science Centre, Harefield Hospital, Harefield, UK*

TERRY C. FORD • *Department of Biology, University of Essex, Colchester, UK*

ANNE-CATHERINE FRICAUD • *Department of Biochemistry, University of Sussex, Brighton, UK*

JOHN M. GRAHAM • *Merseyside Innovation Centre, Liverpool, UK*

NORMAN A. GREGSON • *Department of Anatomy and Cell Biology, United Medical and Dental Schools of Guy's and St. Thomas's Hospitals, London, UK*

ANN L. HUBBARD • *Department of Cell Biology and Anatomy, The Johns Hopkins School of Medicine, Baltimore, MD*

ANTONY P. JACKSON • *Department of Biochemistry, University of Cambridge, UK*

J. PAUL LUZIO • *Department of Clinical Biochemistry, University of Cambridge, Addenbrooke's Hospital, Cambridge, UK*

MALCOLM LYON • *Cancer Research Campaign, Department of Medical Oncology, Christie Hospital and University of Manchester, UK*

ANTHONY L. MOORE • *Department of Biochemistry, University of Sussex, Brighton, UK*

ROBERT K. POOLE • *Microbial Physiology Research Group, Division of Biosphere Sciences, King's College London, UK*

PETER J. RICHARDSON • *Department of Pharmacology, University of Cambridge, UK*

MICHAEL J. SCHELL • *Department of Cell Biology and Anatomy, The Johns Hopkins School of Medicine, Baltimore, MD*

LAURA SCOTT • *Department of Cell Biology and Anatomy, The Johns Hopkins School of Medicine, Baltimore, MD*

JEREMY E. TURNBULL • *Department of Clinical Research and Cancer Research Campaign, Department of Medical Oncology, University of Manchester, Christie Hospital and Holt Radium Institute, Manchester, UK*

ANDREW J. WALTERS • *Department of Biochemistry, University of Sussex, Brighton, UK*

DAVID G. WHITEHOUSE • *Department of Biochemistry, University of Sussex, Brighton, UK*

DAVID J. WINTERBOURNE • *Department of Surgery, St. George's Hospital Medical School, London, UK*

CHAPTER 1

# The Identification of Subcellular Fractions from Mammalian Cells

## John M. Graham

### 1. Introduction
### *1.1. Background*

The chapters in the two volumes of *Biomembrane Protocols* are broadly devoted to the methodologies involved in the structural and functional characterization of cell membranes. In many cases, a study of a particular membrane presupposes that it can be isolated by disruption of the cells and fractionation of the resulting homogenate, and then identified unambiguously. The early chapters of this volume, *Biomembrane Protocols: I. Isolation and Analysis*, are concerned with this fractionation process. Later chapters are concerned with the determination of composition, whereas in the second volume, *Biomembrane Protocols: II. Architecture and Function,* the methods associated with architecture and function predominate.

This chapter is concerned with the problem of membrane identification, so that purity and yield can be established. At least with animal cells—and this chapter will address the problem of identification only in relation to such cells—the most common means of identification of a membrane is an enzymic one. Not only are enzymes valuable markers, but they are also, to some extent, a measure of the functional integrity of the preparation. Enzymic characterization of the membranes of animal cells has been carried out largely on organized tissues, such as liver, intestine, or kidney. In these tissues, the assignation of an enzymic activity to a particular membrane can often be rationalized

From: *Methods in Molecular Biology, Vol. 19: Biomembrane Protocols: I. Isolation and Analysis*
Edited by: J. M. Graham and J. A. Higgins Copyright ©1993 Humana Press Inc., Totowa, NJ

on the basis of the known functions of the tissue. Confirmatory evidence has come from cytochemical localization studies at the electron microscope level.

## 1.2. Mitochondria, Lysosomes, Peroxisomes, and Golgi Membranes

By and large, the internal organelles can be identified by enzymes that are associated with well-established functions. Mitochondria, for example, can be detected by any enzyme associated with electron transport: Normally succinate is used as the substrate and the reduction of the natural electron acceptor, cytochrome c, or an artificial acceptor, such as *p*-iodonitrotetrazolium violet (INT), followed spectrophotometrically. If the mitochondria are to be used for functional studies, it is probably more relevant to use the natural acceptor; if it is only required to assess mitochondrial content of a fraction, then INT is the preferred acceptor since the assay procedure is simpler. Although the mitochondria from such tissues as liver exhibit high activities of this enzyme, very often those from tissue culture cells have much lower actvities and are sometimes difficult to measure (*see* Chapter 9).

Other organelles also have distinct markers: acid phosphatase, aryl sulfatase, and β-galactosidase for lysosomes; catalase for peroxisomes; and sugar transferases (galactosyl transferase is the most commonly used) for Golgi membranes.

## 1.3. Endoplasmic Reticulum

The endoplasmic reticulum from a tissue, such as a rat liver, has a number of enzymic markers, NADPH-cytochrome c reductase and glucose-6-phosphatase being the most widely used. Kidney is the only other tissue in which glucose-6-phosphatase has a clear functional role; reports of its use as an endoplasmic reticulum marker in other cells (particularly tissue culture cells) are not entirely convincing. Indeed there is no clear reason why it should be present in such cells; its function in liver (and kidney) is the control of blood glucose levels. Unless it is clearly insensitive to either fluoride or EDTA (inhibitors of nonspecific acid and alkaline phosphatases, respectively), it should not be used as an endoplasmic reticulum marker.

NADH-cytochrome c reductase has also been used as an endoplasmic reticulum marker; the presence of mitochondria may, however, confuse the results, although the mitochondrial activity is normally inhibited by rotenone.

### *1.4. Plasma Membrane*

In the plasma membranes of organized tissues, such as rat liver, intestine, or kidney, there is clear evidence for functional polarization. The brush border of enterocytes, for example, is associated with carbohydrate digestion, and disaccharidases, such as maltase, have been used as markers. There are, however, a number of markers that are more generally associated with a well-defined brush border membrane. Alkaline phosphatase, leucine-aminopeptidase, and 5'-nucleotidase, for example, tend to be concentrated in such membranes whether they be in enterocytes, hepatocytes, or kidney cells. On the other hand, in the plasma membrane at the opposite pole of the cell, sometimes called the basolateral membrane, the $Na^+/K^+$-ATPase is usually concentrated. In the hepatocyte, the situation is complicated by the division of the basolateral membrane into the contiguous membrane (that part of the plasma membrane next to an adjacent cell) and the sinusoidal membrane (the part next to the blood sinusoid)—in this case, the $Na^+/K^+$-ATPase is most concentrated in the contiguous membrane. Other enzymes, such as hormone-sensitive adenylate cyclase, are associated with areas of the plasma membrane adjacent to the blood supply. Assays for this enzyme are discussed in Chapter 15 of *Biomembrane Protocols: II. Architecture and Function.*

Other chemical markers for the plasma membrane have occasionally been used, e.g., high cholesterol (*see* Chapter 16) or sialic acid (*see* Chapters 25 and 26), but they have been used only after identification has already been achieved enzymatically. Surface antigens are good markers for the plasma membrane, but again they have been used mainly secondarily to enzymes.

## 2. 5'-Nucleotidase

For general comments, *see* Notes 1–5.

### *2.1. Materials*

#### *2.1.1. Measurement by Phosphate Release (see Notes 6 and 7)*
1. Plastic assay tubes (2 mL).
2. Reaction mixture: $0.2M$ Tris-HCl, pH 8.5, 20 m$M$ AMP, 20 m$M$ MgCl$_2$ (*see* Note 8).

3. 10% (w/v) Ascorbic acid.
4. 0.42% (w/v) Ammonium molybdate in $1N$ $H_2SO_4$.
5. Spectrophotometer (visible wavelength).

### 2.1.2. Measurement of Radiolabeled Adenosine, Separated from AMP on an Ion Exchange Column (see Note 9)

1. Glycine-NaOH buffer, pH 8.5: $0.1M$ and $1.0M$.
2. Reaction mixture: $0.1M$ glycine buffer, $0.1M$ $MgCl_2$, and 30 m$M$ [U-$^3$H]-5' AMP (*see* Note 10).
3. Washed anion exchange resin (e.g., Bio-Rad, Richmond, CA, AG 1-X2, 200–400 mesh) stored in 5 vol of $0.1M$ glycine-NaOH, pH 8.5, at 4°C.
4. Pasteur pipets plugged with glass wool. For convenience, attach a small length of silicone rubber tubing to the tip of each pipet, and use a spring clip to clamp it.
5. Plastic assay tubes (1–2 mL).

### 2.1.3. Measurement of Radiolabeled Adenosine by Adsorption of AMP to Barium Sulfate (see Note 11)

1. Plastic assay tubes (2 mL).
2. Low-speed refrigerated centrifuge with appropriate swing-out rotor.
3. Reaction mixture: 100 m$M$ Tris-HCl, pH 8.0, 1 m$M$ $MgCl_2$, 10 m$M$ β-glycerophosphate, 0.25 m$M$ AMP containing tracer amount of [U-$^{14}$C] AMP (4 kBq/mL).
4. $0.3N$ $ZnSO_4$.
5. $0.3N$ $Ba(OH)_2$ (*see* Note 12).

## 2.2. Methods

### 2.2.1. Measurement by Phosphate Release

1. Transfer sample (50–200 μg protein) to enzyme assay tubes, and make up each to 0.5 mL with water.
2. Mix 1 part of ascorbic acid with 6 parts of molybdate, and set up two phosphate assay tubes per enzyme assay, containing 0.7 mL of this mixture.
3. Add 0.5 mL of reaction mixture to the enzyme assay tubes, and incubate at 37°C.
4. At zero time and after 20 min incubation, transfer 0.3 mL of the enzyme assay mixture to a phosphate assay tube.
5. Incubate phosphate assay tubes at 45°C for 20 min, and read the absorbance at 820 nm (*see* Notes 13–15).

### 2.2.2. Measurement of Radiolabeled Adenosine, Separated from AMP by Ion Exchange Column

1. Set up the required number of ion exchange columns (0.5 in of resin in each pipet), and wash with 5 mL of 0.1$M$ glycine buffer. Use the spring clip to keep a small head of buffer above the column.
2. To 0.1 mL of reaction mixture, add the membrane suspension and make up to 1.0 mL with water (*see* Note 16).
3. Incubate at 37°C for 15 min, and then place on ice.
4. Transfer 0.5 mL of assay mixture to each column, and wash through with 4.5 mL of 1.0$M$ glycine buffer.
5. Collect the entire effluent in one tube, and using an appropriate scintillation cocktail, count an aliquot (e.g., 0.5 mL) of the effluent (*see* Note 17).

### 2.2.3. Measurement of Radiolabeled Adenosine by Adsorption to Barium Sulfate

1. Set up the assay tubes in ice: Add 0.5 mL of reaction mixture to each tube; sample and water to 1.0 mL (*see* Note 16).
2. Incubate at 37°C for 30 min.
3. Transfer the tubes to ice, add 0.3 mL $ZnSO_4$ and 0.3 mL $Ba(OH)_2$, and vortex mix.
4. Stand the tubes at 4°C for 20 min with occasional shaking, and then centrifuge at 700$g$ for 15 min.
5. Determine the counts in a 0.5-mL aliquot of the supernatant (*see* Note 17), using a suitable scintillation cocktail (*see* Note 17).

## 3. Ouabain-Sensitive Na+/K+-ATPase

For general comments, *see* Notes 1–4 and 19.

### 3.1. Materials

#### 3.1.1. Measurement by Phosphate Release (see Note 20)

1. Reaction mixture:
   A: 3 m$M$ $MgCl_2$, 3 m$M$ ATP (Tris salt), 130 m$M$ NaCl, 20 m$M$ KCl, 30 m$M$ Histidine buffer, pH 7.5;
   B: Same as A plus 1 m$M$ ouabain.
2. 50% (w/v) Trichloracetic acid (TCA)
3. 5% (w/v) Sodium dodecyl sulfate (SDS).
4. 0.1% (w/v) 2,4 Diaminophenol dihydrochloride in 1% (w/v) sodium sulfite.
5. 0.1% (w/v) Ammonium molybdate in 1$M$ $H_2SO_4$.
6. 5-mL Plastic assay tubes.

### 3.1.2. Using [γ³²P]-Labeled ATP (see Note 21)

1. Reaction mixture: 2 m$M$ ATP containing a tracer amount of [γ³²P]ATP (7 kBq/mL), 0.3$M$ NaCl, 0.06$M$ KCl, 2 m$M$ MgCl$_2$, 40 m$M$ Tris-HCl, pH 7.4.
2. 10 m$M$ Ouabain.
3. Charcoal mixture: 4% (v/v) Norit A (Sigma, St. Louis, MO) 0.1$M$ HCl, 1 m$M$ sodium phosphate, 1 m$M$ sodium pyrophosphate, 0.2 mg/mL bovine serum albumin (*see* Note 22).
4. Wash solution: 0.01$M$ HCl, 1 m$M$ sodium phosphate.
5. Millipore (Bedford, MA) microanalysis filter holder.
6. 2.5-cm Glass-fiber filter disks (Whatman GF/C, Maidstone, UK).
7. 5-mL Plastic assay tubes (with caps).

## 3.2. Methods

### 3.2.1. Measurement by Phosphate Release

1. For each sample, set up separate tubes containing 1 mL of reaction mixture A or B, and bring to 37°C (approx 5 min).
2. Add 25 µL of sample to each pair of tubes, and incubate at 37°C for 5–10 min (*see* Notes 7, 13–15, and 23).
3. Add 0.1 mL of TCA, and put on ice for about 10 min.
4. Bring to room temperature; add 1 mL each of SDS, diaminophenol, and molybdate; mix well after each addition (*see* Note 24).
5. After 20 min, read the absorbance at 660 nm.

### 3.2.2. Using [γ³²P]-Labeled ATP

1. The ATPase activity of each sample is measured in the absence and presence of ouabain. Add 0.5 mL of reaction mixture to all the assay tubes on ice, and add 10 µL of ouabain to half of them.
2. Add sample and make up the volume to 1.0 mL with water. Make sure that the final concentration of any gradient compound, such as sucrose, is <0.2$M$.
3. Set up additional tubes for determining the total counts and the background counts (*see* Note 16).
4. Incubate at 37°C for 30 min.
5. Place the tubes in ice; add 3.0 mL ice-cold charcoal mixture, and leave at 4°C for 30 min with occasional shaking. The unhydrolyzed ATP is adsorbed on to the charcoal (*see* Note 25).
6. Filter the suspension through a glass-fiber filter disk in the Millipore filter holder directly into a scintillation vial. Wash the residue twice with 3.0 mL of the wash solution (*see* Notes 26 and 27).
7. Use wide window settings on a scintillation counter to measure the ³²P$_i$ by Cerenkov radiation.

## 4. Alkaline Phosphatase
## (and Acid Phosphatase)

For general comments, *see* Notes 1–4.

### *4.1. Materials*

1. Substrate: 16 m*M* *p*-nitrophenyl phosphate (aq.).
2. Buffers: (alkaline) 50 m*M* sodium borate-NaOH, pH 9.8; (acid) 180 m*M* sodium acetate-acetic acid, pH 5.0.
3. 1.0*M* MgCl$_2$.
4. 0.25*M* NaOH.
5. 1.5-mL Eppendorf tubes.

### *4.2. Method*

1. Make a reaction mixture of the appropriate pH by mixing 5 mL substrate and 5 mL buffer. For alkaline phosphatase only, add 20 µL of the MgCl$_2$.
2. To 0.2 mL of the reaction mixtures, add 50 µL of sample and incubate at 37°C for 30 min. Set up a control with 0.2 mL of buffer.
3. Stop the reaction by adding 0.6 mL of 0.25*M* NaOH, and remove any particulate material by centrifugation in a microfuge for 1–2 min.
4. Read the absorbance of the supernatant against the control at 410 nm (*see* Note 29).

## 5. Aminopeptidase

For general comments, *see* Notes 1–4.

### *5.1. Materials*

1. Substrate: 25 m*M* leucine *p*-nitroanilide.
2. 50 m*M* Phosphate buffer, pH 7.2.
3. Recording spectrophotometer (*see* Note 28).

### *5.2. Method*

1. In a cuvet, mix 1.0 mL buffer with 0.1 mL sample.
2. When the absorbance at 405 nm is steady, add 0.1 mL of substrate, and record the absorbance so that a measurable linear increase is obtained (*see* Note 29).

## 6. β-Galactosidase

For general comments, *see* Notes 1–4.

### *6.1. Materials*

1. 0.05*M* Citrate-phosphate buffer, pH 4.3.
2. Reaction mixture: 6 m*M* *p*-nitrophenyl β-D-galactopyranoside, 0.5% (w/v) Triton X100, 0.05*M* citrate-phosphate buffer, pH 4.3.

3. 0.25*M* Glycine-NaOH, pH 10.
4. 1.5-mL Eppendorf tubes.

## 6.2. Method

1. Centrifuge an appropriate volume of the sample in an assay tube, and resuspend the pellet in 0.5 mL of the reaction mixture. For a control, add 0.5 mL buffer to an identical pellet (*see* Note 30).
2. Incubate at 37°C for 30 min.
3. Stop the reaction by addition of 1.0 mL of the glycine-NaOH, and remove sedimentable material in a microfuge for 1–2 min.
4. Read the absorbance of the supernatant at 410 nm against the control (*see* Note 29).

# 7. Aryl Sulfatase

For general comments, *see* Notes 1–4.

## 7.1. Materials (see Note 31)

1. 0.1*M* Sodium acetate buffer, pH 5.6.
2. 30 m*M* Lead acetate (**care: very toxic**).
3. 5.0 m*M* 4-Methylumbelliferyl sulfate (4-MUS), freshly made up.
4. 0.2*M* Glycine-carbonate buffer, pH 10.3, containing 1.0*M* EDTA.
5. 5.0 m*M* 4-Methylumbelliferone (4-MU) in the glycine buffer, freshly made up.
6. 1.5-mL Plastic assay tubes.
7. Spectrophotofluorimeter.

## 7.2. Method

1. Make up a reaction mixture containing buffer, lead acetate, 4-MUS, and water in the ratio of 10:2:2:5 (v/v/v/v).
2. To 0.19 mL of reaction mixture, add 10 µL of sample (about 10 µg protein) and incubate at 37°C for 15 min.
3. Stop the reaction with 1.0 mL of glycine buffer.
4. Measure the fluorescence of the solution (emission wavelength 470 nm), using an excitation wavelength of 387 nm.
5. Calibrate the instrument using various dilutions of the 4-MU standard (*see* Note 32).

# 8. Succinate Dehydrogenase

For general comments, *see* Notes 1–4.

## 8.1. Materials

### 8.1.1. Using Cytochrome c (see Note 33)

1. Reaction mixture: 50 m*M* sodium phosphate buffer, pH 7.4, 2 m*M* potassium cyanide, 0.625 mg/mL cytochrome c (*see* Notes 34 and 35).
2. 660 m*M* Sodium succinate in the phosphate buffer (*see* Note 35).
3. Recording spectrophotometer (full-scale absorbance set to 0.2) with 1-mL glass cuvet (*see* Note 28).

### 8.1.2. Using INT (see Note 36)

1. Buffer: 0.05*M* phosphate, pH 7.5.
2. Substrate medium: 0.01*M* sodium succinate in buffer.
3. Acceptor (INT): *p*-iodonitrotetrazolium violet (2.5 mg/mL) in buffer.
4. Ethyl acetate:ethanol:trichloracetic acid (EET), 5:5:1 (v/v/w).
5. 1.5-mL Eppendorf assay tubes.
6. Microfuge.

## 8.2. Methods

### 8.2.1. Using Cytochrome c

1. Place 1.0 mL of reaction mixture in a glass cuvet; add 10–50 μL of sample and mix well.
2. Record the absorbance at 550 nm for about 2 min or until a steady baseline is obtained (*see* Notes 37 and 38).
3. Add 0.1 mL of succinate; mix well and continue to record the absorbance until a satisfactory and linear increase in $A_{550}$ has been recorded (*see* Note 39).

### 8.2.2. Using INT

1. Pellet the membrane from an appropriate volume of suspension in an assay tube (*see* Note 30).
2. Resuspend the pellet in 0.3 mL of substrate medium, and incubate at 37°C for 10 min. For a control, resuspend an identical pellet in 0.3 mL of buffer.
3. Add 0.1 mL of INT, and continue the incubation for a further 10 min.
4. Stop the reaction with 1 mL of EET, and after spinning down any precipitate in a microfuge for 2 min, measure absorbance at 490 nm against the control (*see* Note 40).

## 9. Catalase

For general comments, *see* Notes 1–4.

## 9.1. Materials (see Note 41)

1. 0.01*M* Phosphate buffer, pH 7.0.
2. Substrate: 6 m*M* hydrogen peroxide in 0.01*M* phosphate, pH 7.0.
3. 6*N* $H_2SO_4$.
4. 0.01*N* $KMnO_4$.
5. Membrane samples in 0.25*M* sucrose in the phosphate buffer.
6. 1.5-mL Eppendorf assay tubes.

## 9.2. Method (see Note 42)

Carry out all operations at 4°C.

1. Treat the test samples in the following manner: To 50 µL of sample add 0.5 mL of $H_2O_2$; vortex mix, and after 3 min, add 0.1 mL of $H_2SO_4$.
2. Set up a blank for each sample as in step 1, but add the acid to the sample first.
3. In addition, set up a reagent blank: 0.1 mL of $H_2SO_4$, 0.5 mL of peroxide, and 50 µL of buffer and a standard: 0.55 mL of buffer and 0.1 mL of $H_2SO_4$.
4. The spectrophotometer blank contains 0.7 mL of water, 0.1 mL of $H_2SO_4$, and 0.55 mL of buffer.
5. Add 0.7 mL of $KMnO_4$ to all the tubes set up in (1)–(3) and read the absorbance at 480 nm within 1 min against (4) (*see* Note 43).

# 10. Galactosyl Transferase

For general comments, *see* Notes 1–4 and 44.

## 10.1. Materials (see Note 45)

### 10.1.1. Using N-Acetylglucosamine as Acceptor

1. Basic reaction mixture: Mix 300 m*M* of $MnCl_2$, 30 m*M* of mercaptoethanol, 0.04*M* HEPES-NaOH buffer, pH 7.0, and UDP-galactose (1 mg/mL) containing a trace amount of UDP-[6-$^3$H]-galactose (70 kBq) in the ratio of 1:1:2:1 (v/v/v/v).
2. Divide the basic reaction mixture into two portions; to one add 1/5 vol of water (assay mixture A), and to the other add 1/5 vol of 0.4*M* N-acetylglucosamine (assay mixture B).
3. 0.3*M* EDTA.
4. Suspension of Dowex-1-Cl⁻ anion exchange resin (e.g., Sigma Product Number 1X2-200) in water.
5. 1.5-mL Plastic assay tubes.
6. Pasteur pipets (short form) plugged with glass wool fitted with 1–2 cm of silicone tubing.
7. Scintillation cocktail (*see* Note 17).

### 10.1.2. Using Ovalbumin as Acceptor

1. Buffer: 0.1$M$ cacodylate-HCl, pH 6.2.
2. Incubation medium: 10 m$M$ ATP, 10 m$M$ MnCl$_2$, 30 m$M$ β-mercapto-ethanol, 0.2% (v/v) Triton X-100, 5% (w/v) ovalbumin in buffer made up within 1 h of use.
3. 12.5 m$M$ UDP-galactose containing a trace amount of UDP-[6-$^3$H]-galactose (70 kBq).
4. 1-mL Plastic assay tubes.
5. Filter paper disks (approx 2.4 cm).
6. 10% (w/v) Trichloroacetic acid (TCA).
7. Pins and a polystyrene board.
8. Scintillation cocktail.

## 10.2. Methods

### 10.2.1. Using N-Acetylglucosamine as Acceptor

1. Incubate each sample (10 μL) with 60 μL of both assay mixtures at 37°C for 30 min (*see* Notes 16 and 47).
2. Meanwhile, set up the appropriate number of columns of Dowex in the Pasteur pipets (approx 2 cm bed height); wash with 3 vol of water, and clip the silicone tubing to keep a small head of water (approx 2 mm) above the column.
3. Stop the enzyme reaction with 20 μL of EDTA.
4. Apply all the incubation mixture to the ion exchange resin, and then wash each column with 2 × 0.5 mL water (*see* Note 46).
5. Collect the eluate into a scintillation vial; add a suitable scintillation cocktail (*see* Note 17) and count.

### 10.2.2. Using Ovalbumin as Acceptor

1. Prepare at least two filter paper disks per sample, and number them lightly with a pencil.
2. Pin the disks to the polystyrene board so that they are held well above the board surface.
3. Mix 50 μL of incubation medium with 50 μL of membrane suspension, add 10 μL of UDP-galactose, and incubate at 37°C (*see* Note 48).
4. At zero time and after 10 min, withdraw 50-μL fractions and spot onto filter paper disks (*see* Note 49).
5. Transfer the disks to ice-cold TCA, and leave for several hours or overnight (*see* Note 50).
6. Wash the disks several times in distilled water; dry; transfer to a scintillation vial and count using a routine scintillant.

## 11. NADPH-Cytochrome c Reductase

For general comments, *see* Notes 1–4.

### *11.1. Materials (see Note 51)*

1. 50 m$M$ Phosphate buffer, pH 7.7.
2. NADPH (2 mg/mL) and cytochrome c (25 mg/mL), both made up freshly in the phosphate buffer (*see* Note 52).
3. 10 m$M$ EDTA.
4. Recording spectrophotometer with 1-mL glass cuvet (*see* Note 28).

### *11.2. Method*

1. Make up an assay mixture containing buffer, cytochrome c, and EDTA in the ratio 90:5:1 (v/v/v).
2. In a cuvet, mix 1 mL of the assay mixture with 10–20 μL of sample (*see* Note 53), and record the absorbance at 550 nm (full-scale deflection equivalent to 0.2 absorbance units).
3. When the absorbance is steady, add 0.1 mL of NADPH, and record the absorbance until a measurable, steady increase in value is achieved (*see* Notes 39 and 54).

## 12. Glucose-6-Phosphatase

For general comments, *see* Notes 1–4.

### *12.1. Materials (see Note 55)*

1. Assay solution: Mix 0.1$M$ glucose-6-phosphate, sodium salt; 35 m$M$ histidine buffer, pH 6.5, and 10 m$M$ EDTA in the ratio 2.5:1:1 (v/v/v). Also prepare a control solution of buffer and EDTA (3.5:1).
2. 8% (w/v) Trichloroacetic acid (TCA).
3. 2.5% (w/v) Ammonium molybdate in 2.5$M$ H$_2$SO$_4$.
4. Reducing solution: 0. 5g 1-amino-2-naphthol-4-sulfonic acid, 6.0 g anhydrous Na$_2$SO$_3$ dissolved in 200 mL of 15% NaHSO$_3$ (*see* Note 56).
5. 4-mL Plastic assay tubes.
6. Bench-top refrigerated centrifuge.
7. Boiling water bath.

### *12.2. Method*

1. To 0.45 mL of assay solution or control solution, add 50 μL of sample and incubate at 37°C for 30 min. Also set up a reagent blank containing 50 μL of buffer in place of sample.
2. Stop the reaction by adding 2.5 mL of ice-cold TCA, and keep at 0–4°C for 20 min.

3. Centrifuge at 1000$g$ for 15 min (preferably at 4°C).
4. Without disturbing the pellet, remove 1.0 mL of supernatant, and add 1.15 mL of water, 0.25 mL of molybdate, and 0.1 mL of reducing solution, mixing well after each addition.
5. Heat tubes in boiling water bath for 10 min.
6. Allow the tubes to cool, and read absorbance at 820 nm (*see* Notes 7 and 57).

## 13. Notes

1. Identification of membranes by enzyme assay and any analysis of membrane proteins can be hampered by the presence of proteases in the homogenate. These degradative enzymes may be derived from a number of sources (e.g., lysosome breakdown) that are normally compartmentalized in the intact cell or tissue. It is common to include a cocktail of proteases at all stages during the homogenization and fractionation procedures to minimize this problem. One mixture routinely used consists of pepstatin, leupeptin, chymostatin, antipain, and aprotonin. A stock solution contains 5.0 mg of each (except pepstatin, 0.5 mg) in 10 mL of water and use at 1:100 in all solutions. Phenylmethylsulfonyl fluoride (PMSF), a general protease inhibitor, may also be used at about 0.1 m$M$ (10 m$M$ stock in isopropanol).
2. With large numbers of samples, always set up the assay tubes in ice, particularly if the incubation time is short, and return them, if necessary, to ice after incubation prior to further processing.
3. The amount of membrane material needed in the assays (mg protein/mL) will vary with the sensitivity of the method and the purity of the membrane preparation. Generally speaking, to avoid conditions where the concentration of the substrate may become rate-limiting, in any assay, the amount of substrate utilized should be not more than 10–15% of the total.
4. If the sample is suspended in 0.25$M$ sucrose buffered with a low concentration (5–10 m$M$) of an organic buffer, then there should be no interference with accurate measurement of the enzyme activity. If the sample is taken directly from a sucrose gradient, then the final concentration of sucrose in the assay mixture should be <0.2$M$.
5. The methods and the use of this marker are discussed by Glastris and Pfeiffer *(1)*, Graham *(2)*, and Widnell *(3)*. Virtually all the reported methods are carried out in an alkaline Tris or glycine buffer in the presence of Mg$^{2+}$ and the substrate. Alkaline phosphatase may interfere with the assay: There are two ways to overcome this. If the AMP is radiolabeled, then a second unlabeled substrate, such as β-glycerophosphate, in the incubation mixture can serve as the substrate for the alkaline phosphatase, or a specific inhibitor of the alkaline phosphatase, such as levamisole (50 μ$M$), can be included.

6. This assay is adapted from Widnell *(3)*.
7. Many enzymes that are used as markers have been assayed by determination of the release of inorganic phosphate from the substrate. As with all phosphate estimations, scrupulous attention must be paid to the reduction of background levels from a variety of sources. Radioisotope assays avoid these problems and have improved sensitivity.
8. The enzyme reaction mixture can be kept frozen for at least 3 mo.
9. This assay is adapted from Glastris and Pfeiffer *(1)*.
10. Any commercially available AMP, labeled in any of the carbon or hydrogen atoms, can be used.
11. This assay is adapted from Avruch and Wallach *(4)*.
12. Barium hydroxide solutions are difficult to prepare from solid because of the rapid formation of insoluble barium carbonate in the presence of atmospheric carbon dioxide. It is preferable to use ready-made solutions of both $ZnSO_4$ and $Ba(OH)_2$, which are available commercially from Sigma.
13. Direct measurement of the absorbance may be difficult at the high end of the protein range, in which case extract the colored product into 1 mL of isoamyl alcohol and read at 795 nm (but *see* Note 14).
14. If you use isoamyl alcohol, check that the plastic tubes are resistant to this solvent; also check the wavelength of the absorbance maximum of the colored product in the solvent.
15. With larger amounts of protein (in excess of 200 µg), it may be necessary to precipitate it using TCA prior to the phosphate colorimetric assay (*see* Sections 3. and 12.).
16. In radioisotope assays, always include two additional types of sample: Blank (without membrane) that is treated identically with the sample tubes to give the background hydrolysis and a blank that is not processed to determine the total number of counts in each tube.
17. The amount of effluent that can be counted will depend on the nature of the cocktail, its quenching characteristics, its water miscibility, and its tolerance of ionized molecules.
18. When removing the supernatant for scintillation counting, take care not to disturb the barium sulfate pellet, which may not be well packed. Also there is a tendency for the sulfate to adhere to the walls of the tube, so make sure that the pipet tip withdraws a sample from the center, not from close to the wall.
19. The $Na^+/K^+$–ATPase has been measured by a number of techniques. The ability of ouabain to inhibit selectively this enzyme means that assays are commonly carried out in the presence and absence of the inhibitor, the difference in activity being that attributable to the $Na^+/K^+$–ATPase.

The concentrations of ouabain used are usually of the order of 1 m*M*. Attention must be paid to the concentrations of Na$^+$ and K$^+$: The total concentration of both ions should be optimally 150 m*M*, and the ratio Na$^+$ to K$^+$ about 6.5 *(5)*. Two techniques are described for direct measurement of the amount of ATP hydrolyzed. There is also a coupled assay in which the released ADP is used to generate pyruvate in the presence of phosphoenolpyruvate and pyruvate kinase; the amount of pyruvate is then determined by the loss of NADH (reduction in 340 nm absorbance) in the presence of lactate dehydrogenase. The method requires a thermostatted cuvet compartment in a recording spectrophotometer *(6)*.

20. Assay is adapted from Esmann *(5)*.
21. Assay is adapted from Avruch and Wallach *(4)*.
22. When making up the charcoal mixture, first mix together all the components, except for the Norit A and the bovine serum albumin. Then add the Norit from a measuring cylinder, and finally add the bovine serum albumin in the form of a concentrated solution. Stir well before storing at 4°C.
23. The assay samples should contain between 60 and 300 nmol inorganic phosphate. Standards should be assayed of the same range.
24. A more sensitive phosphate assay method (10–100 nmol phosphate) can be used *(5)*. The two reagents are (A) 0.75 g ascorbic acid in 25 mL 0.5*M* HCl, with 1.25 mL 10% (w/v) ammonium molybdate added with stirring and (B) a solution of 2% (w/v) sodium-m-arsenite, 2% (w/v) trisodium citrate (dihydrate), and 2% (v/v) acetic acid. Add 0.5 mL of A to the assay and cool to 4°C. After 6 min, bring to 37°C, and add 1.5 mL of B. After 10 min, read at 850 nm.
25. When adding the charcoal mixture to the assay tubes, keep it well stirred and use a 10-mL glass pipet with a wide-bore tip to dispense it.
26. The most convenient way of filtering the suspension is to house the filter unit in a tall-form scintered glass funnel whose stem is connected to a vacuum line. The stem of the Millipore filter unit leads into a scintillation vial, which is supported on the glass frit of the funnel.
27. Do not allow the charcoal residue to dry out before rinsing with the wash solution.
28. With a double-beam spectrophotometer, replace the substrate with buffer in the reference cuvet.
29. The molar extinction coefficient of the product, *p*-nitrophenol, is 9620.
30. The centrifugation conditions needed for this step will depend on the material being processed. This method is ideal for material that can be pelleted at about 15,000 rpm in a microfuge for 10 min. If you have to use a membrane suspension rather than a pellet, then do not add more than 50 µL to the substrate medium.

31. Assay adapted form Chang et al. *(7)*.
32. This is an extremely sensitive assay. The actual sensitivity depends on the spectrophotofluorimeter, and the range of concentrations of 4-MU used to calibrate the machine will vary. Amounts of 4-MUS and membrane protein may also be adjusted to allow for equipment sensitivity. Always check that the enzyme concentration is the rate-limiting factor.
33. Assay adapted from Mackler et al. *(8)*.
34. Cyanide is very poisonous. Always handle cyanide in a fume cupboard, and dispense the solid into a preweighed capped vial. Before adding the cytochrome c to the reaction mixture, check that the pH of the phosphate buffer/cyanide mixture is not >7.8. Crystalline cyanide normally contains alkali to prevent the formation of gaseous hydrogen cyanide when in contact with water. If the pH is >7.8, reduce it very cautiously by slow addition of 50 m$M$ sodium dihydrogen phosphate. Always have available the appropriate antidotes to cyanide poisoning (against both cyanide in solution and hydrogen cyanide gas). **Consult your safety officer before using cyanide.**
35. The buffer and succinate can be stored frozen, but the cytochrome c solution should be made up freshly. Cyanide solutions can be kept frozen for periods up to 1 mo. Do not use cyanide solutions that exhibit a yellow/brown tinge.
36. Assay adapted from Davis and Bloom *(9)*.
37. The incubation of the mitochondria in the absence of substrate is necessary for two reasons. A hypotonic buffer is used to permeabilize the membranes so the cytochrome c can gain access to its binding site on the inner membrane. In this medium, the mitochondria will swell, and any absorbance fluctuation attributable to light scattering changes by the mitochondria must be allowed to occur before substrate is added.
38. Some workers prefer to add the cyanide separately after this initial swelling phase.
39. The molar extinction coefficient of reduced cytochrome c is 27,000.
40. The molar extinction coefficient of reduced INT is 19,300.
41. Method adapted from Cohen et al. *(10)*.
42. The timing of this assay is critical. The most convenient way of adding the $KMnO_4$ is immediately prior to measuring the absorbance. Remember that the samples are ice-cold and condensation on the cuvet may be a problem.
43. The rationale of the assay is that the $KMnO_4$ is used to determine the unused peroxide.
44. Assays for galactosyl transferase use an acceptor glycoprotein, such as fetuin, that needs to be desialylated and degalactosylated before use

*(11)*; ovalbumin that can be used without modification *(12)* or a simple molecule, such as *N*-acetylglucosamine *(13)*, that forms the disaccharide *N*-acetyllactosamine. For ease of assay, only the ovalbumin and *N*-acetylglucosamine methods are given.

45. Assay adapted from Fleischer et al. *(13)*.
46. The resin can be easily regenerated by stirring with an excess of 1*M* HCl for 2–3 h, then washing with large volumes of distilled water in a scintered glass funnel until the filtrate is neutral and storing dry.
47. Crude fractions and homogenates may cause significant hydrolysis of the UDP-galactose rather than transfer of the galactose to its artificial acceptor, *N*-acetylglucosamine. This will be detected in assay mixture A. Only the difference in counts between assay mixtures A and B represents activity attributable to the transferase enzyme.
48. The amount of product is linear up to approx 500 µg microsomal protein.
49. Use preliminary experiments to determine the optimum time (up to 20 min): If you use more than two time-points, reduce the size of sample as appropriate.
50. Large numbers of filter disks can be treated with TCA by simply dropping them into 250–350 mL of TCA in a 500-mL beaker, using gentle agitation to keep them suspended.
51. Method adapted from Williams and Kamin *(14)*.
52. Both the NADPH and cytochrome c should be kept on ice, and the NADPH should be protected from the light.
53. If the volume of sample needs to be increased to more than 50 µL, make the appropriate reduction to the volume of buffer.
54. The membrane suspension needs to be well dispersed to avoid changes in absorbance attributable to light scattering by particles moving within the beam of incident radiation. With turbid suspensions, it is permissible to use a mild detergent, such as deoxycholate (approx 1%) to clarify.
55. Adapted from Aronson and Touster *(15)*.
56. Ideally, the reducing solution should be prepared freshly each time. It is, however, permissible to keep the reagent in a dark bottle at 4°C. A precipitate often forms with keeping, which is difficult to redissolve. When decanting, take care to leave any solid in the bottle. The dry components of the reducing solution can be purchased as a ready prepared mixture (Fiske Subbarow Reducer) from Sigma or made up from individual reagents.
57. Although the phosphate assay given here is the one used by a number of workers, the alternative methods given in Sections 2.1.1. and 3.1.1. can be tried.

# References

1. Glastris, B. and Pfieffer, S. B. (1974) Mammalian membrane marker enzymes: sensitive assay for 5'-nucleotidase and assay for mammalian 2'3'-cyclic nucleotide 3'-phosphohydrolase. *Methods Enzymol.* **32,** 124–131.
2. Graham, J. M. (1975) Cellular membrane fractionation, in *New Techniques in Biophysics and Cell Biology* (Pain, R. and Smith, B. J., eds.), Wiley, London, pp. 1–42.
3. Widnell, C. C. (1974) Purification of rat liver 5'-nucleotidase as a complex with sphingomyelin. *Methods Enzymol.* **32,** 368–374.
4. Avruch, J. and Wallach, D. F. H. (1972) Preparation and properties of plasma membrane and endoplasmic reticulum fragments from isolated rat fat cells. *Biochim. Biophys. Acta* **233,** 334–347.
5. Esmann M. (1988) ATPase and phosphatase activity of Na$^+$/K$^+$-ATPase: molar and specific activity, protein determination. *Methods Enzymol.* **156,** 105–115.
6. Norby, J. G. (1988) Coupled assay of Na$^+$/K$^+$-ATPase activity. *Methods Enzymol.* **156,** 116–119.
7. Chang, P. L., Rosa, N. E., and Davidson, R. G. (1981) Differential assay of arylsulfatase A and B activities: a sensitive method for cultured human cells. *Anal. Biochem.* **117,** 382–389.
8. Mackler, B., Collip, P. S., Duncan, H. M., Rao, N. A., and Huennekens, F. H. (1962) An electron transport particle from yeast: purification and properties. *J. Biol. Chem.* **237,** 2968–2974.
9. Davis, G. A. and Bloom, F. E. (1973) Subcellular particles separated through a histochemical reaction. *Anal. Biochem.* **51,** 429–435.
10. Cohen, G., Dembiec, D., and Marcus, J. (1970) Measurement of catalase activity in tissue extracts. *Anal. Biochem.* **34,** 30–38.
11. Kim, Y. S., Perdomo, J., and Nordberg, J. (1971) Glycoprotein biosynthesis in small intestinal mucosa. I A study of glycosyl transferases in microsomal subfractions. *J. Biol. Chem.* **246,** 5466–5476.
12. Beaufay, H., Amar-Costesec, A., Feytmans, E., Thines-Sempoux, D., Wibo, M., Robbi, M., and Berthet, J. (1974) Analytical study of microsomes and isolated subcellular fraction from rat liver. *J. Cell Biol.* **61,** 188–200.
13. Fleischer, B., Fleischer, S., and Ozawa, H. (1969) Isolation and characterization of Golgi membranes from bovine liver. *J. Cell Biol.* **43,** 59–79.
14. Williams, C. H. and Kamin, H. (1962) Microsomal triphosphopyridine nucleotide-cytochrome c reductase of liver. *J. Biol. Chem.* **237,** 587–595.
15. Aronson, N. N. and Touster, O. (1974) Isolation of rat liver plasma membrane fragments in isotonic sucrose. *Methods Enzymol.* **31,** 90–102.

CHAPTER 2

# The Isolation of Nuclei and Nuclear Membranes from Rat Liver

## John M. Graham

## 1. Introduction

### 1.1. Preparation of Nuclei from Rat Liver

Of all the organelles in rat hepatocytes, the nucleus is the largest and the most dense. It is therefore relatively easy to isolate in high purity and high yield. The majority of methods *(1,2)* involve the homogenization of the liver in isoosmotic sucrose followed by differential centrifugation and isopycnic purification through a sucrose density barrier. Because of their high density (>1.30 g/cm$^3$), the sucrose density barrier concentration is about 2.25$M$ or 60% (w/w) and, consequently, very hyperosmotic. This is one of the disadvantages of the method in that the nuclei collapse significantly during their passage through the sucrose barrier. They are subsequently resuspended in an isoosmotic medium in which they swell. The hydration of the DNA within the nucleus therefore changes during the purification process. Moreover, the high viscosity of the barrier causes a considerable reduction in the rate of sedimentation of the nuclei, which is the reason for the use of high relative centrifugal forces (rcf) (normally 100,000$g$). Furthermore, high viscosity sucrose solutions are difficult to prepare and measure out accurately.

The concentration of the sucrose density barrier and, consequently, the rcf needed can be reduced (to 0.88$M$ and 2000$g$, respectively), in which case the separation of the nuclei will be on the basis of sedimentation rate rather than density. The nuclei are less stressed by the

From: *Methods in Molecular Biology, Vol. 19: Biomembrane Protocols: I. Isolation and Analysis*
Edited by: J. M. Graham and J. A. Higgins Copyright ©1993 Humana Press Inc., Totowa, NJ

isolation conditions, but contamination by other organelles tends to be higher. This can be reduced by inclusion of 0.5% Triton X-100 in the homogenization medium—this will partially solubilize all membranous structures of the hepatocyte except the nuclear membrane, which is relatively stable to nonionic detergents.

It is customary to include a number of ions in the homogenization medium *(1,2)* and in all the subsequent barrier or gradient solutions to stabilize the nuclei. Divalent cations $Ca^{2+}$ and $Mg^{2+}$ at concentrations of 1–5 m$M$ and $K^+$ at 10–25 m$M$ are common. This poses problems if the isolation of other organelles from the homogenate is desired: Mitochondria, e.g., are best isolated, in the presence of EDTA, so it is preferable to commit one part of the liver to the preparation of nuclei and another part to the preparation of other organelles.

To avoid the problems that solutions of high concentrations of sucrose pose in the isopycnic separation of nuclei, one of the new iodinated density gradient media may be used. Nycodenz™ solutions, e.g., can provide the appropriate high density for purification of nuclei without the corresponding high osmolarity or viscosity of sucrose solutions *(3)*. The nuclei therefore do not suffer the same osmotic stress, and the rcf required can be substantially reduced.

The nuclei can be purified either from the whole homogenate or from a 1000-g pellet. A purification method for each source, one that uses sucrose and one that uses Nycodenz, is given. Which method is chosen depends on operational requirements and on the volume of homogenate. Large volumes of homogenate may become inconvenient to handle, in which case the use of a concentrated pellet may be preferable. The two purification methods in sucrose and Nycodenz are adapted from Blobel and Potter *(1)* and Graham et al. *(3)*, respectively.

## 1.2. Purification of Nuclear Membranes

The isolation of the nuclear membrane (or envelope) from purified nuclei has been achieved using contrasting conditions: low ionic strength *(4,5)* or high ionic strength *(6)*. Low-ionic-strength preparations are generally easier to carry out, but the yields tend to be rather variable, and residual DNA can be problematical. High-ionic-strength preparations are more complex in their execution, but they can yield very pure preparations.

There are some important considerations to the preparation of nuclear envelopes. During the DNA extraction procedures, proteolysis can occur, so protease inhibitors, such as phenylmethylsulfonyl fluoride (PMSF), are included *(4)*. However, the efficacy of low-ionic-strength procedures seems to rely on some degree of proteolysis, since inclusion of PMSF impairs the separation. A major problem can be the oxidative crosslinking of intranuclear and membrane proteins, which is minimized by the inclusion of β-mercaptoethanol *(4)*. Finally, for reasons that are not entirely clear, the presence of $Ca^{2+}$ seems to be deleterious to envelope isolation; consequently the inclusion of EGTA at an early stage in the preparation has been recommended *(4)*.

Three methods are given: a high-ionic-strength preparation based on the method of Kaufmann et al. *(6)* and two low-strength preparations based on the methods of Harris and Milne *(7)*, as modified by Comerford et al. *(4)*, and Kay et al. *(5)*.

## 2. Materials

### 2.1. Preparation of Homogenate and Crude Nuclear Pellet from Rat Liver

1. TKM: 10 m*M* Tris-HCl, pH 7.6, 25 m*M* KCl, and 5 m*M* $MgCl_2$. The Tris can be replaced by any organic buffer such as HEPES or Tricine.
2. Homogenization medium: 0.25*M* sucrose in TKM.
3. Potter-Elvehjem homogenizer (glass/Teflon™) with a clearance of about 0.07 mm, with thyristor-controlled electric motor. A 20-mL capacity model is suitable for one or two livers.
4. 50-mL Conical-bottomed polypropylene centrifuge tubes.
5. Cheesecloth or nylon mesh (pore size 75 μm).
6. 20-mL Syringe fitted with metal cannula (1 mm id).

### 2.2. Purification of Nuclei from Homogenate by Sucrose Density Barrier Centrifugation

1. TKM (*see* Section 2.1., item 1).
2. 0.25*M* and 2.3*M* (61% [w/w]) sucrose in TKM.
3. Polycarbonate tubes (size depends on sample volume) for an ultracentrifuge swing-out rotor.
4. Refractometer.
5. 20-mL Syringe.

## 2.3. Purification of Nuclear Pellet in a Nycodenz Gradient

1. Make a stock solution of 60% (w/v) Nycodenz in TKM (*see* Section 2.1., item 1), and prepare solutions of 20, 30, 35, 40, and 50% by dilution with TKM.
2. Loose-fitting Dounce homogenizer (about 5-mL capacity).
3. Polycarbonate tubes of about 14-mL capacity to fit a high-speed or ultra-centrifuge swing-out rotor.
4. Refractometer.

## 2.4. Purification of Nuclear Membrane in High-Ionic-Strength Medium

1. Buffer 1: 0.25$M$ sucrose, 5 m$M$ MgSO$_4$, 50 m$M$ Tris-HCl, pH 7.5, 1 m$M$ PMSF, 1 m$M$ EGTA.
2. Buffer 2: 0.2 m$M$ MgSO$_4$, 10 m$M$ Tris-HCl, pH 7.5, 1 m$M$ PMSF, 1 m$M$ EGTA.
3. 2$M$ NaCl in Buffer 2
4. β-Mercaptoethanol.
5. DNase I (2.5 mg/mL).
6. RNase A (2.5 mg/mL).
7. Refrigerated low-speed centrifuge.

## 2.5. Purification of Nuclear Membrane in Low-Ionic-Strength Media

### 2.5.1. Using 1 mM NaHCO$_3$

1. 1 m$M$ NaHCO$_3$.
2. Gradient solutions: 0.25$M$, 1.5$M$ (43% [w/w]), 1.8$M$ (50% [w/w]), and 2.0$M$ (55% [w/w]) sucrose in 10 m$M$ Tris-HCl, pH 7.4.
3. DNase I (10 μg/mL) in 1 m$M$ NaHCO$_3$.
4. Syringe and metal cannula (about 1 mm id).

### 2.5.2. Using a Tris Buffer

1. Wash medium: 0.25$M$ sucrose, 1 m$M$ MgCl$_2$ adjusted to pH 7.4 with NaHCO$_3$.
2. DNase I (100 μg/mL) must be freshly prepared.
3. Digestion medium: 10% (w/v) sucrose, 0.1 m$M$ MgCl$_2$, 10 m$M$ Tris-HCl, 5 m$M$ β-mercaptoethanol. Adjust one half to pH 7.4, and the other to pH 8.5.
4. Glass rod.

# 3. Methods

## 3.1. Preparation of Homogenate and Crude Nuclear Pellet from Rat Liver

Carry out all operations at 4°C, and make sure that the homogenizers are precooled. All g-forces are $g_{av}$.

1. Kill a 250 g rat by cervical dislocation; open the abdominal and thoracic cavities, and displace the alimentary canal to expose the hepatic portal vein.
2. Cut the major blood vessels above the liver, and inject 50 mL of ice-cold homogenization medium into the hepatic portal vein from a syringe, as close to the liver as possible. The lobes of the liver should appear a creamy brown as the blood is washed out of the vasculature (*see* Note 1).
3. Rapidly excise the liver, transfer to a chilled beaker, and chop very finely with scissors until the minced pieces are no more than a couple of millimeters thick.
4. If the liver has not been perfused, remove some of the blood by swirling the minced liver in homogenization medium; then allow the pieces to settle out before decanting the liquid.
5. Transfer about one-third of the mince to the glass vessel of the homogenizer. Pour on about 10 mL of homogenization medium, and disrupt the liver with no more than eight up-and-down strokes of the pestle rotating at about 500 rpm (*see* Notes 2–6).
6. Check that you have achieved at least 95% cell breakage under the phase contrast microscope.
7. Transfer the homogenate to a chilled beaker, and repeat the procedure until all the liver mince has been processed.
8. Strain the homogenate through either two to three layers of cheesecloth or nylon gauze (75 μm) to remove any clumps of cells and connective tissue. This suspension can be used directly for further purification (*see* Sections 2.2. and 3.2.).
9. Alternatively, place the filtered homogenate in a chilled 50-mL plastic conical-bottomed centrifuge tube, and centrifuge at 1000g for 5 min in a swing-out rotor.
10. Remove the supernatant, either by decantation or using a syringe and metal cannula (about 1 mm id). If you decant it, take care for the pellet may slide in the tube.
11. Add about 2 mL of homogenization medium to the pellet, and resuspend it by vortex mixing (*see* Note 7).

### 3.2. Purification of Nuclei from Homogenate by Sucrose Density Barrier Centrifugation

1. Add double the volume of $2.3M$ sucrose in TKM to the homogenate to bring the sucrose concentration to $1.62M$. Because of the difficulty of adding accurate volumes of a very viscous solution, check the sucrose concentration by measuring the refractive index (it should be 1.411).
2. Transfer the homogenate to the tubes of an appropriately sized swingout rotor, e.g., 28 mL/tube in a $6 \times 38.5$ mL rotor, and underlay with $2.3M$ sucrose in TKM to fill the tube.
3. Centrifuge at approx $100,000g$ for 30 min.
4. Remove all the material from above the pellet by decanting: To remove all traces of material from the walls of the tube, with the tube inverted, carefully wash the walls with TKM from a syringe.
5. Resuspend the pellet in a suitable medium (*see* Notes 8–11).

### 3.3. Purification of Nuclei from a Nuclear Pellet in a Nycodenz Gradient

1. Dislodge the pellet from step 9 (*see* Section 3.1.) into about 3 mL of 35% Nycodenz in TKM, using a glass rod; transfer to a loose-fitting Dounce homogenizer, and resuspend the pellet using five to six gentle strokes of the pestle.
2. Check the concentration of Nycodenz with a refractometer, and adjust to 35% (refractive index 1.3956), if necessary, by the dropwise addition of 60% Nycodenz in TKM.
3. Make up the following discontinuous gradient of Nycodenz: 1.0-mL 50%, 2.0-mL 40%, 4-mL 35%, 3-mL 30%, and 2-mL 20%.
4. Centrifuge at $13,000g$ for 1 h.
5. Using a syringe attached to a metal cannula (about 1 mm id), remove the sharp band at the 40/50% Nycodenz interface.
6. Dilute with an equal volume of TKM, and harvest the purified nuclei by centrifugation at $1000g$ for 20 min (*see* Notes 8–11).

### 3.4. Purification of Nuclear Membrane in High-Ionic-Strength Medium

Carry out all steps at 4°C.

1. Purify the nuclei as described in Sections 3.2. or 3.3., with 1 m$M$ EGTA and 1 m$M$ PMSF in all the solutions.
2. Resuspend the purified nuclei in Buffer 1 (approx $5 \times 10^8$/mL); add 1/10 vol of each of the enzyme solutions, and incubate for 1 h.

3. Centrifuge at 1000*g* for 10 min, and resuspend in the same volume of Buffer 2.
4. With gentle stirring, add the 2*M* NaCl dropwise so that the final concentration is 1.6*M* and β-mercaptoethanol to 1% (v/v).
5. After incubating for 15 min, centrifuge at 2000*g* for 30 min.
6. Again resuspend in Buffer 2, and make 1.6*M* with respect to NaCl.
7. Repeat step 4, and resuspend the pellet of purified nuclear envelopes in a suitable buffer (*see* Notes 12–14).

### 3.5. Low-Ionic-Strength Preparation
#### 3.5.1. Using 1 mM NaHCO₃

1. Resuspend the nuclei (prepared from one rat liver as in Sections 3.2. or 3.3., using 1 m*M* EGTA in all media) in 10 mL of 1 m*M* NaHCO$_3$, pH 7.4, containing 1 m*M* EGTA and 5 m*M* *N*-ethylmaleimide by shaking.
2. After 5 min, centrifuge at 31,500*g* for 5 min.
3. Resuspend the pellet in the same medium by gently aspirating into and ejecting from a syringe and wide-bore metal cannula.
4. Repeat step 2.
5. Disperse the pellet in 10 mL of DNase solution, and incubate at 22°C for 15 min.
6. Centrifuge the nuclear envelopes at 31,500*g* for 5 min, and wash them four times in 10 mL of 1 m*M* NaHCO$_3$.
7. Remove any aggregated material by centrifuging at 500*g* for 5 min.
8. If you want to purify the membranes further, suspend them in 1–2 mL of 1 m*M* NaHCO$_3$; layer over a discontinuous gradient comprising 3 mL each of the gradient solutions; and centrifuge at 100,000*g* for 90 min. The membranes band at the 1.5/1.8*M* sucrose interface (*see* Notes 12–14).

#### 3.5.2. Using a Tris Buffer

1. Suspend the nuclei from one rat liver, prepared as described in Sections 3.2. or 3.3. (using 1 m*M* EGTA in the isolation medium), in 50 mL of wash medium and centrifuge at 700*g* for 5 min.
2. After pouring off the supernatant, wipe the walls of the tube to remove residual medium.
3. Add 3 mL of 0.1 m*M* MgCl$_2$: Use a glass rod to resuspend the pellet.
4. Then add 0.2 mL of the DNase solution, and immediately add 12 mL of digestion medium, pH 8.5; vortex and incubate for 15 min at 22°C.
5. Add 15 mL of ice-cold water; mix and centrifuge at 38,000*g* for 15 min.
6. After decanting the supernatant, repeat steps 3–5 using digestion medium, pH 7.4.
7. Resuspend the pellet of nuclear envelopes in an appropriate buffer (*see* Notes 12–14).

# 4. Notes

1. Make sure that the tip of the perfusion cannula does not reach into one of the branches of the portal vessel; otherwise only selected lobes will be blanched.
2. Unless you are dealing with large numbers of livers, the homogenization is most easily carried out in batches of about 4 g of liver.
3. Always try to use the smallest volume homogenizer that is feasible; homogenize large amounts of material in four or five batches in a smaller homogenizer rather than processing all the material in a much larger one. The larger the volume of homogenizer, the more difficult it is to reproduce the exact conditions.
4. The number of strokes needed will depend on a number of parameters (the size, clearance, thrust, and speed of the pestle and the size of the liver fragments).
5. Tissue, such as muscle, will need a blender type of homogenizer rather than a liquid shear one.
6. The homogenization methodology for tissue culture cells needs to be worked out for individual cell types (*see* Chapter 9). The inclusion of both monovalent and divalent cations in the homogenization medium usually makes the rupture of cultured animal cells by standard liquid shear techniques (Potter-Elvehjem or Dounce) very difficult. It is preferable to devise a homogenization strategy that will cause relatively rapid and efficient cell breakage, and then to add cations to the homogenate as soon as possible so as to protect the nuclei.
7. Nuclei from cultured cells tend to be more fragile than those from intact tissues and need to be handled with great care to prevent damage.
8. There are significant advantages in using the method in Section 3.3. for the purification of nuclei. Because the nuclei are not exposed to such hyperosmotic conditions in Nycodenz, they are less dense than they are in sucrose. Consequently, they can be banded at an interface rather than sedimented into a pellet. Pellet formation at high rcfs should be avoided wherever it is practical, since the aggregation that occurs is often difficult to reverse and the shear forces generated by resuspension (usually involving passage through the narrow bore of a pipet, metal cannula, or syringe needle) frequently lead to breakdown of the material.
9. Removal of dense sucrose by dilution with buffer also exposes the nuclei to a sudden change in osmolarity: This is much less marked with iodinated gradient media. The lower viscosity of Nycodenz also means that a 1:1 dilution with buffer provides a medium through which the nuclei can be harvested rapidly by sedimentation at 1000$g$.

10. The purity and integrity of the nuclear preparations can best be checked by phase contrast microscopy at about 400× magnification. The nuclei should appear as distinct dark gray bodies with no amorphous material surrounding them.

11. Estimation of the DNA *(8)* gives information about the recovery of the nuclei and, if expressed per unit of protein, will give some indication of the purification. The only enzyme that is relatively easy to measure, which is considered to be characteristic of the nuclear membrane, is nucleoside triphosphatase *(9)*. The presence of enzymes characteristic of other organelles can be used to detect contaminants, but enzymes of the smooth endoplasmic reticulum tend to be present since the outer nuclear membrane is continuous with the smooth endoplasmic reticulum.

12. Preparations should be checked for DNA, RNA, protein, and phospholipid content. Typical values for the percentage recovery of DNA are 1–4% for the low-ionic-strength methods and about 0.5% for the high-ionic-strength method. Phospholipid recoveries are generally about 40–50%.

13. Activities of nucleoside triphosphatase tend to be higher in the low ionic strength preparations.

14. For studies into nucleocytoplasmic transport, it is often more satisfactory to use sealed nuclear envelope vesicles. The method of Riedel and Fasold *(10)* is often used. The method involves incubation of the nuclei in heparin in a Tris-HCl-phosphate buffer; filtration through scrubbed nylon fiber; recovery of material from the filtrate by centrifugation; and washing the pellet in sucrose/Tris-HCl/KCl/MgCl$_2$/CaCl$_2$ buffer. The reader is referred to the original paper for full details.

## References

1. Blobel, G. and Potter, V. R. (1966) Nuclei from rat liver: isolation method that combines purity with high yield. *Science* **154,** 1662–1665.
2. Harris, J. R. and Agutter, P. S. (1970) A negative staining study of human erythrocyte ghosts and rat liver nuclear membrane. *J. Ultrastr. Res.* **33,** 219–232.
3. Graham, J. M., Ford, T., and Rickwood, D. (1990) Isolation of the major subcellular organelles from mouse liver using Nycodenz gradients without the use of an ultracentrifuge. *Anal. Biochem.* **187,** 318–323.
4. Comerford, S. A., McLuckie, I. F., Gorman, M., Scott, K. A., and Agutter, P. S. (1985) The isolation of nuclear envelopes: effects of thiol group oxidation and of calcium ions. *Biochem. J.* **226,** 95–103.
5. Kay, R. R., Fraser, D., and Johnston, I. R. (1972) A method for the rapid isolation of nuclear membranes from rat liver. *Eur. J. Biochem.* **30,** 145–154.
6. Kaufmann, S. H., Gibson, W., and Shaper, J. H. (1983) Characterization of the major polypeptides of the rat liver nuclear envelope. *J. Biol. Chem.* **258,** 2710–2719.

7. Harris, J. R. and Milne, J. F. (1974) A rapid procedure for the isolation and purification of rat liver nuclear envelope. *Biochem. Soc. Trans.* **2,** 1251–1253.
8. Peters, D. L. and Dahmus, M. E. (1979) A method of DNA quantitation for localization of DNA in metrizamide gradients. *Anal. Biochem.* **93,** 306–311.
9. Agutter, P. S., Cockrill, J. B., Lavine, J. E., McCaldin, B., and Sim, R. B. (1979) Properties of mammalian nuclear envelope nucleoside triphosphatase. *Biochem. J.* **181,** 647–658.
10. Riedel, N. and Fasold, H. (1987) Preparation and characterization of nuclear envelope vesicles from rat liver nuclei. *Biochem. J.* **241,** 203–212.

# CHAPTER 3

# Isolation of Mitochondria, Mitochondrial Membranes, Lysosomes, Peroxisomes, and Golgi Membranes from Rat Liver

## John M. Graham

## 1. Introduction

### 1.1. The "Heavy" Mitochondrial Fraction

This fraction is defined broadly as the material that will sediment at about 3500g for 10 min from a postnuclear supernatant. It contains the nuclei that failed to sediment at 1000g for 5–10 min, the largest mitochondria, and very few other organelles, such as lysosomes and peroxisomes.

Functional studies, such as those concerned with respiratory control and the production of ATP, demand a rapid preparation time and the use of the mildest of procedures. Since respiration depends critically on the permeability properties of the inner mitochondrial membrane, exposure of the organelles to the osmotic stresses associated with purification in sucrose density gradients or to the hydrostatic pressures of a high centrifugal field should be avoided. Moreover, the enzymic apparatus involved in the processes are unique to mitochondria, so the presence of low-degree contamination by lysosomes or peroxisomes may be acceptable, as long as these organelles, particularly the lysosomes, remain intact. The "heavy fraction" is therefore often used for these studies (*see* Sections 2.1. and 3.1.). It is also used for the fractionation of inner and outer mitochondrial membranes.

From: *Methods in Molecular Biology, Vol. 19: Biomembrane Protocols: I. Isolation and Analysis*
Edited by: J. M. Graham and J. A. Higgins Copyright ©1993 Humana Press Inc., Totowa, NJ

Some of the "heavy" mitochondria will sediment with the nuclear pellet (1000*g*), and this pellet can be used to obtain a purified mitochondrial fraction, essentially devoid of lysosomal or peroxisomal contamination by a simple sucrose barrier technique (*see* Sections 2.2. and 3.2.).

## 1.2. Resolution of Inner and Outer Mitochondrial Membranes

Because the outer membrane of the mitochondrion has a higher concentration of cholesterol than does the inner, an agent such as digitonin, which complexes cholesterol, tends to cause disruption of the outer membrane, while leaving the inner relatively intact. Most methods are based on that of Schnaitman and Greenawalt *(1)*, and there are many variations in details of medium composition and so forth. Once the outer membrane has been removed, the residual inner membrane and matrix are sometimes called the mitoplast. Osmotic shock of the mitoplasts, followed by sonication, leads to the production of inner membrane vesicles. The method described herein employs modifications made by Hackenbrock and Hammon *(2)* and Coty et al. *(3)*. Outer membrane preparations usually rely on a controlled swelling procedure based on the method of Parsons et al. *(4)*.

## 1.3. Resolution of Lysosomes, Peroxisomes, and Golgi Membranes from the "Light" Mitochondrial Fraction

The "light" mitochondrial fraction is defined broadly as the material that sediments from the "heavy" mitochondrial supernatant at relative centrifugal forces (rcf) that vary from 10,000–16,000*g* for 10 min. This material contains the remaining mitochondria together with lysosomes, peroxisomes, and Golgi membranes. Golgi membranes will only sediment under these conditions if the homogenization is gentle so that they do not vesiculate. Generally speaking, this is only feasible in cells containing a large well-defined Golgi stack (e.g., hepatocytes). The poorly organized Golgi membranes present in many tissue culture cells tend to vesiculate under the more severe homogenization conditions required to disrupt these cells.

Depending on the operational requirements, the postnuclear super-natant may be centrifuged at 6000$g$ for 10 min, rather than 3000$g$, to eliminate relatively more of the mitochondria; alternatively, the "heavy" pellet step may be omitted entirely if all the mitochondria need to be harvested as well.

Traditionally, both the lysosomes and the Golgi have been pre-pared from a "light" mitochondrial fraction by sucrose gradient centri-fugation *(5,6)*. The density of the Golgi membranes is lower than that of any other component that sediments at 6000–15,000$g$ for 10 min, so in buoyant density separations, they always band on top of all others and are relatively easy to isolate *(7,8)*. In sucrose density gradients, the lysosomes tend to band at a slightly higher density (1.18–1.22 g/cm$^3$) than do the mitochondria (1.16–1.21 g/cm$^3$), but there is consider-able overlap. These banding densities mean that sucrose gradients in the range 30–60% (w/v) are often used for their purification.

The disadvantages of using such gradients are that resolution, par-ticularly that of mitochondria, lysosomes, and peroxisomes, is never satisfactory and the osmotic activity of the high sucrose concentrations causes shrinkage of organelles. The viscosity of the sucrose gradient has three consequences: (1) the centrifugation conditions normally demand an ultracentrifuge; (2) the high rcfs or long centrifugation times used increase the disruption of organelles, particularly mito-chondria; and (3) the harvested gradient fractions need dilution at least threefold with buffer to reduce the concentration of sucrose to a level that will permit rapid pelleting of the organelles for subsequent analysis. Wattiaux and his colleagues *(9,10)* have used the iodinated density gradient compound metrizamide for the separation of mito-chondria, lysosomes, and peroxisomes. The lower osmolarities and viscosities of metrizamide solution (compared to sucrose solutions of similar densities) improved resolution of the major organelles.

The method described in Sections 2.5. and 3.5. is designed to sepa-rate Golgi membranes, mitochondria, and lysosomes from rat liver using Nycodenz™ (11) rather than metrizamide. A "total" mitochon-drial fraction, which combines both heavy and light, will be used. The operator can use this fraction or either of its components. Alter-native methods using both sucrose and Percoll gradients for resolv-ing Golgi and lysosomes from cultured cells are given in Chapter 5.

# 2. Materials

## 2.1. "Heavy" Mitochondrial Fraction by Differential Centrifugation

1. A 20-mL Potter-Elvehjem homogenizer (clearance 0.07 mm) attached to an overhead electric (thyristor-controlled) motor.
2. A loose-fitting Dounce homogenizer (30 mL).
3. 20-mL Syringe fitted with a wide-bore metal cannula (1–2 mm id).
4. 50-mL Conical-bottomed polypropylene tubes for a refrigerated low-speed centrifuge (swing-out rotor); 50-mL polycarbonate tubes for a refrigerated high-speed centrifuge (fixed-angle rotor).
5. Mannitol buffer: $0.2M$ Mannitol, $0.05M$ sucrose, 10 m$M$ KCl, 1 m$M$ EDTA, 10 m$M$ HEPES-NaOH, pH 7.4.
6. Glass rod.

## 2.2. "Heavy" Mitochondrial Fraction by Density Barrier

1. Potter-Elvehjem homogenizer (*see* Section 2.1., item 1).
2. Dounce homogenizer (*see* Section 2.1., item 2).
3. Centrifuge tubes (*see* Section 2.1., item 4).
4. $0.25M$ Sucrose, 10 m$M$ HEPES-NaOH, pH 7.5.
5. $0.25M$ Sucrose, 10 m$M$ HEPES-NaOH, pH 7.5, 1 m$M$ MgCl$_2$.
6. $2.4M$ (63% [w/w]) Sucrose, 10 m$M$ HEPES-NaOH, pH 7.5, 1 m$M$ MgCl$_2$.
7. Cheesecloth or nylon mesh (75 nm pore size).

## 2.3. Inner Mitochondrial Membranes

1. Isolation medium: 220 m$M$ mannitol, 70 m$M$ sucrose, 20 m$M$ HEPES-NaOH, pH 7.4, 0.5 mg/mL fatty-acid-free bovine serum albumin (BSA).
2. Mitochondria produced by standard differential centrifugation technique (Sections 2.1. and 3.1.) in the isolation medium.
3. 2% Digitonin in isolation medium: Prepare this by dissolving the digitonin in the medium (without added BSA) with heating; then add the BSA.
4. Loose-fitting Dounce homogenizer.
5. Sonicator (*see* Note 13).
6. Clear plastic (e.g., polycarbonate) tubes for fixed-angle rotors of high-speed and ultracentrifuges. Volumes will depend on amount of material.

## 2.4. Outer Mitochondrial Membranes

1. 20-m$M$ Phosphate buffer, pH 7.2, 0.02% BSA.
2. Clear plastic (e.g., polycarbonate) tubes (about 50 mL) for fixed-angle rotors for low-speed and high-speed centrifuges.

3. 0.73$M$ (23% [w/w]), 1.12$M$ (33% [w/w]), and 1.48$M$ (43% [w/w])
sucrose in the phosphate buffer.
4. Swing-out rotor (about 5-mL capacity tubes) for ultracentrifuge.

## 2.5. Nycodenz Gradient Fractionation of the "Total" Mitochondrial Fraction

1. Potter-Elvehjem homogenizer (*see* Section 2.1., item 1).
2. Dounce homogenizer (*see* Section 2.1., item 2).
3. Syringe and cannula (*see* Section 2.1., item 3).
4. Centrifuge tubes for fixed-angle rotors (*see* Section 2.1., item 4).
5. Buffer: 1 m$M$ EDTA, 10 m$M$ HEPES-NaOH, pH 7.6.
6. Homogenization medium (HM): 0.25$M$ sucrose in buffer.
7. Nycodenz solution of 10, 15, 20, 23, 40, and 50% (w/v) in buffer.
8. Thin-walled polycarbonate (or polyallomer) tubes for a swing-out rotor capable of spinning at 20,000 rpm (about 50,000$g$) in a suitable centrifuge (*see* Notes 1 and 2).
9. A refrigerated microfuge (approx 15,000 rpm) with a fixed-angle rotor that accepts tubes of at least 1.5 mL.
10. Refractometer.
11. Gradient unloader (upward displacement)—optional.

## 3. Methods

All g-forces are $g_{av}$.

### 3.1. "Heavy" Mitochondrial Fraction by Differential Centrifugation

1. Precool all the apparatus to 0–4°C.
2. Rapidly excise the liver from a young adult male Wistar rat, killed by cervical dislocation; place in a beaker and chop finely with scissors.
3. Suspend the crudely minced tissue in about 30 mL of mannitol buffer, and transfer half the material to the glass vessel of the Potter-Elvehjem homogenizer. Homogenize using four to six up-and-down strokes of the pestle rotating at about 500 rpm (*see* Note 3).
4. Repeat the process with the other half of the liver; combine the homogenates, and make up to about 50 mL with buffer (*see* Note 4).
5. Distribute the homogenate between two polypropylene tubes, and centrifuge at 1000$g$ for 5 min.
6. Remove the supernatant by aspiration into a 20-mL syringe (fitted with cannula). The pellet tends to be bipartite. The lower larger layer contains nuclei, red blood cells, and cell debris; on the top of this is a thin brown layer of the most rapidly sedimenting mitochondria, so aspirate the supernatant quite close to the pellet.

7. Distribute the collected supernatants between two polycarbonate tubes, and centrifuge at 3500*g* for 10 min (*see* Note 5).

8. Remove the tubes from the rotor carefully to avoid disturbing the pellet or the layer of lipid floating at the surface. Aspirate the supernatant, keeping the tip of the Pasteur pipet (*see* Note 6) in the meniscus so that the floating lipid layer is also removed. Also remove the loose-packed pinkish layer that overlies the brown "heavy" mitochondria (*see* Notes 7 and 8).

9. Using a tissue, wipe away any remaining lipid that is adhering to the wall of the tube. The removal of this lipid is crucial to the recovery of good respiratory coupling.

10. Add about 10 mL of the mannitol buffer to each tube, and gently resuspend the pellet using a round-ended glass rod. Try not to disturb the very dark button at the wall of the tube: This contains red blood cells and aggregated organelles, which are difficult to resuspend.

11. Transfer the crude suspension to the Dounce homogenizer, and complete the resuspension with two or three gentle strokes of the pestle. Then add more buffer to regain the original volume, and recentrifuge in new polycarbonate tubes at 5000*g* for 10 min (*see* Note 4).

12. Repeat steps 7–9 at least twice, and suspend the final pellet in about 10 mL of mannitol buffer.

### 3.2. "Heavy" Mitochondrial Fraction by Density Barrier (see Note 9)

Carry out all operations at 0–4°C.

1. Produce a rat liver homogenate as in Section 3.1., steps 1–4, using the sucrose/HEPES homogenization medium, and filter it using either three layers of cheesecloth or the nylon mesh. Do not force the homogenate through the filter; aid filtration only by agitation with a glass rod.

2. Centrifuge the filtered homogenate at 1000*g* for 10 min.

3. After removing the supernatant, gently pour 15 mL of the 0.25*M* sucrose/1m*M* MgCl$_2$ onto the pellet, and swirl the medium to resuspend it crudely without disturbing the lowermost red layer.

4. Decant the material into the Dounce homogenizer, and complete the resuspension by using three or four gentle strokes of the pestle. Mix in the 2.4*M* sucrose to a final concentration of 1*M* (check that the refractive index is about 1.382).

5. Transfer the suspension to a 50-mL polycarbonate tube, overlay with 10 mL of 0.25*M* sucrose/HEPES, and centrifuge at 35,000*g* for 10 min in a fixed-angle rotor (*see* Note 10). Remove the supernatant.

6. The pellet is bipartite: The upper brown layer contains the mitochondria. From a pipet, run 0.25$M$ sucrose/HEPES down the side of the inclined tube (pellet lowermost); resuspend the upper layer with a gentle swirling action, and decant carefully (*see* Note 11).

### 3.3. Inner Mitochondrial Membranes
### (see Note 12)

Carry out all operations at 0–4°C.

1. Add the digitonin solution to the mitochondrial suspension (50 mg protein/mL) so that the final concentration of digitonin is 0.75%, and stir for 15 min.
2. Dilute with 3 vol of isolation medium, and homogenize the suspension in the Dounce homogenizer using three or four gentle strokes of the pestle.
3. Centrifuge the suspension at 10,000$g$ for 10 min to pellet the mitoplasts.
4. Decant the supernatant, and recentrifuge at the same speed.
5. This supernatant can be centrifuged at 140,000$g$ for 1 h to sediment the remaining material, which is enriched in outer membranes.
6. Resuspend the two pellets of mitoplasts in distilled water (2 mg protein/mL) to lyse them and sediment the inner membrane "ghosts" at 10,000$g$ for 15 min.
7. Inner membrane vesicles are produced by mild sonication of the ghosts using eight cycles of 15 s sonication and 15 s "resting" (*see* Note 13).
8. Remove large particles by centrifugation at 10,000$g$ for 10 min, and recentrifuge the supernatant at 200,000$g$ for 30 min to pellet the vesicles.

### 3.4. Outer Mitochondrial Membranes
### (see Note 14)

Carry out all operations at 0–4°C.

1. Prepare the mitochondria from a single rat liver as described previously (*see* Sections 2.1. and 3.1.).
2. Centrifuge the mitochondria at 10,000$g$ for 10 min; decant the supernatant, and resuspend the pellet in the residual liquid using a glass rod.
3. Over a period of about 3 min, add 70 mL of the phosphate buffer with stirring.
4. Centrifuge at 35,000$g$ for 20 min.
5. Resuspend the pellet in the same volume of phosphate buffer (as in step 3), and recentrifuge at 1900$g$ for 15 min to sediment the inner membrane.
6. Recentrifuge the supernatant at 35,000$g$ for 20 min to pellet the outer membrane, and then resuspend the pellet in about 1.5 mL of buffer.

7. In 5-mL tubes for a swing-out rotor, make a discontinuous gradient of 1.2 mL each of 0.73, 1.12, and 1.48*M* sucrose and 1.2 mL of sample (*see* Note 15). Centrifuge at 115,000*g* for 1 h. The outer membranes band at the 0.73/1.12*M* interface.
8. Recover the banded material using a Pasteur pipet and treat as required.

## *3.5. Nycodenz Gradient Fractionation of a "Total" Mitochondrial Fraction (see Note 16)*

Carry out all operations at 4°C.

1. Produce a postnuclear supernatant from a rat liver homogenate as described in Section 3.1., steps 1–6, using the buffer described in Section 2.5.
2. Distribute the supernatant between two 50-mL polycarbonate tubes, and centrifuge in a fixed-angle rotor at 15,000*g* for 10 min. This may be increased to 20 min if your main interest is the Golgi membranes.
3. Remove as much of the supernatant as possible, and resuspend the pellet(s) in the smallest volume of HM possible: For a 12-g rat liver, the final volume should be no more than 10 mL. Use either two or three strokes of the pestle of a Dounce homogenizer to do this, or use a syringe and metal cannula alternately to direct a jet of HM at the pellet and draw up into the syringe.
4. Mix with an equal volume of 50% Nycodenz, and check that the refractive index is close to and not less than 1.3750; adjust with buffer or 50% Nycodenz if necessary (*see* Note 17).
5. Produce the following discontinuous Nycodenz gradient by underlaying: 10, 15, 20, 23%, sample in 25%, and 40% (in the ratio 2:2:2:3:5:2), and centrifuge the gradients at 52,000*g* for 1.5 h (*see* Note 2).
6. Collect the banded material from the gradient by aspiration into a Pasteur pipet of in equivolume (1 or 2 mL depending on gradient size) fractions using a gradient unloader (upward displacement is the method of choice).
7. To concentrate the material, dilute fractions with half to an equal volume of buffer, and centrifuge in a refrigerated microfuge at about 15,000 rpm for 20 min. Aspirate the supernatant carefully, and resuspend the pellets in HM by vortex mixing.
8. Carry out the appropriate enzyme analysis either on concentrated material or on fractions straight from the gradient (*see* Notes 18–20, *see also* Chapter 1).

## 4. Notes

1. The tube volume will depend on the amount of material being processed: For a 12-g rat liver, tubes of about 38 mL are appropriate; for smaller amounts of material, 16–17 mL tubes.

2. The stated rcf can be generated in an ultracentrifuge swing-out rotor or in one of the new "Supra" high-speed machines. In routine high-speed centrifuges, many swing-out rotors are not capable of this rcf. Lower rcfs can be used; increase the time so that the rcf × time value is correct.

3. Alternative sources of material may require different homogenization procedures. Beef heart, for example, is disrupted using a rotating blades homogenizer.

4. For larger amounts of liver, the volumes of buffer used in steps 1–5 should be scaled up proportionately. For steps 10–12, the volumes can be reduced by up to one-half.

5. Depending on requirements, the rcf used to pellet the mitochondria may be altered. The lowest rcf routinely used is approx 3500g: The higher the rcf used, the greater will be the recovery, but the contamination by other organelles, such as lysosomes and peroxisomes, will also be greater.

6. A Pasteur pipet attached to a vacuum line through a trap is the best mode of aspiration, but a syringe and cannula may be used.

7. The definition between the pinkish rough endoplasmic reticulum and the most poorly packed mitochondria from the first centrifugation at 3500g may not be very clear. To avoid undue loss of mitochondria, this material need only be partially removed at this stage. During the subsequent washings of the mitochondrial pellet, the contaminants tend to remain in the supernatant rather than to pellet and are thus easier to remove.

8. Since the pellet tends to slide as the meniscus passes over it, keep the tip of the pipet or cannula close to the wall of the tube away from the pellet.

9. Method adapted from Fleischer et al. *(12)*.

10. A fixed-angle rotor is used for the barrier separation to facilitate the resuspension of the upper pellet without disturbing the nuclei.

11. If brand-new polycarbonate tubes are used for the barrier separation, the lower part of the pellet (nuclei) may tend to slide off the wall of the tube during resuspension. It is advisable to use tubes whose inner surface has become slightly worn with use.

12. Based on the method of Schnaitman and Greenawalt *(1)*.

13. Because of the variation in sonication equipment, the operator should optimize the sonication conditions by monitoring the vesiculation process. In the first instance, this can be a determination of the amount of material that sediments at the lower rcf compared to that at the higher rcf. The composition of the higher-speed pellet should also be observed by electron microscopy.

14. Based on the method of Parsons *(4)*, Hackenbrock and Hammon *(2)*, and Coty et al. *(3)*.

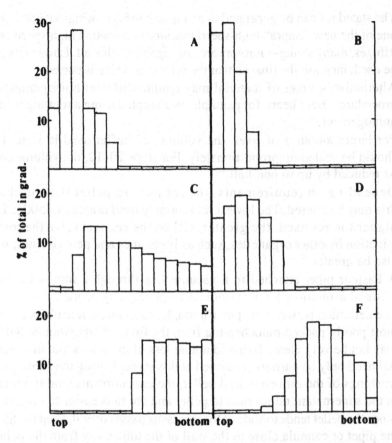

Fig. 1. Fractionation of a rat liver mitochondrial fraction in Nycodenz gradients: Distribution of marker enzymes. Panels **A, C,** and **E**: Rat liver mitochondrial fraction was suspended in 25% (w/v) Nycodenz and a discontinuous gradient made of 2 mL 40% Nycodenz, 5 mL sample, 3 mL 23%, and 2 mL each of 20, 15, and 10% Nycodenz, and centrifuged at 52,000g for 1.5 h. Panels **B, D,** and **F**: Rat liver mitochondrial fraction was suspended in 40% (w/v) Nycodenz and a discontinuous gradient made of 3 mL of sample and 2 mL each of 30, 26, 24, and 19% Nycodenz, and centrifuged at 52,000g for 2.5 h. Panels **A** and **B** are galactosyl transferase; panels **C** and **D** are β-galactosidase, and panels **E** and **F** are succinate-INT reductase.

15. With rotors and samples of different volume, the relative volumes of sample and sucrose solutions should be kept approximately the same. Larger-volume gradients may require longer periods of centrifugation because of the increased sedimentation path length.

16. Method adapted from Graham et al. *(11)*.

17. Because the sample is layered under 23% Nycodenz, it is important to ensure that the concentration of Nycodenz is not <25%. Otherwise lay-

ering will be difficult; the small concentration of sucrose (about 0.12$M$) that is also present will add slightly to its overall density.

18. By varying the gradient profile and/or changing the position at which the sample is loaded, the separation of the various organelles can be modulated. For example, if the crude mitochondrial pellet is resuspended directly in 40% (w/v) Nycodenz and the gradient modified to 19, 24, 26, 30%, and sample (in the ratio 2:2:2:2:3), then the separation of mitochondria and lysosomes is increased, whereas that of the lysosomes and Golgi is diminished.

19. By suspending the sample in 35% Nycodenz and underlayering it with 40% Nycodenz, some resolution of the mitochondria and peroxisomes can be obtained: The peroxisomes tend to band at a slightly higher density than the mitochondria.

20. The distribution of enzymes in two types of Nycodenz gradient separation is shown in Fig. 1.

## References

1. Schnaitman, C. A. and Greenawalt, J. W. (1968) Enzymatic properties of the inner and outer membranes of rat liver mitochondria. *J. Cell Biol.* **38**, 158–175.
2. Hackenbrock, C. R. and Hammon, K. M. (1975) Cytochrome c oxidase in liver mitochondria. Distribution and orientation determined with affinity purified immunoglobulin and ferritin conjugate. *J. Biol. Chem.* **250**, 9185–9197.
3. Coty, W. A., Wehrle, J. P., and Pedersen, P. L. (1979) Measurement of phosphate transport in mitochondria and in inverted inner membrane vesicles of rat liver. *Methods Enzymol.* **56**, 353–359.
4. Parsons, D. F., Williams, G. R., and Chance, B. (1966) Characterization of isolated and purified preparations of outer and inner membranes of mitochondria. *Ann. NY Acad. Sci.* **137**, 643–666.
5. Maunsbach, A. B. (1974) Isolation of kidney lysosomes. *Methods Enzymol.* **31**, 330–339.
6. Graham, J. M. (1984) Isolation of subcellular organelles and membranes, in *Centrifugation—A Practical Approach* (Rickwood, D., ed.), IRL at Oxford University, Oxford, UK, pp. 161–182.
7. Fleischer, B. and Fleischer, S. (1970) Preparation and characterization of Golgi membranes from rat liver. *Biochim. Biophys. Acta* **219**, 301–319.
8. Graham, J. M. and Winterbourne, D. J. (1988) Subcellular localization of the sulphation reaction of heparan sulphate synthesis and transport of the proteoglycan to the cell surface in rat liver. *Biochem. J.* **252**, 437–445.
9. Collot, M., Wattiaux-de-Coninck, S., and Wattiaux, R. (1976) Separation of cell organelles, in *Biological Separations in Iodinated Density Gradient Media* (Rickwood, D., ed.), IRL at Oxford University, Oxford, UK, pp. 119–137.
10. Wattiaux, R., Wattiaux-de-Coninck, S. Ronveaux-Dupal, M. F., and Dubois, F. (1978) Isolation of rat liver lysosomes by isopycnic centrifugation in a metrizamide gradient. *J. Cell Biol.* **78**, 349–368.

11. Graham, J. M., Ford, T., and Rickwood, D. (1979) Isolation of the major subcellular organelles from mouse liver using Nycodenz gradients without the use of an ultracentrifuge. *Anal. Biochem.* **187,** 318–323.
12. Fleischer, S., McIntyre, J. O., and Vidal, J. C. (1979) Large-scale preparation of rat liver mitochondria in high yield. *Methods Enzymol.* **55,** 32–39.

CHAPTER 4

# Continuous-Flow Electrophoresis

*Application to the Isolation of Lysosomes and Endosomes*

## John M. Graham

## 1. Introduction

### 1.1. Background

Continuous-flow electrophoresis (CFE), which separates particles on the basis of surface charge density, should be regarded as an adjunct to centrifugation rather than as an alternative for membrane fractionation. The equipment is expensive and has been used for relatively specialized tasks, not for the routine isolations from a total homogenate.

Although the technique offers some distinct advantages over centrifugation, it has not gained wide popularity for membrane fractionation. Centrifugation can subject particles to considerable hydrostatic pressures, and pelleting can cause an aggregation of particles that is sometimes difficult to reverse. Additionally, some of the media used may cause osmotic stress. In CFE, particles are exposed to unit gravitational field, and the media are normally isoosmotic. Although the imposed voltages may be high (e.g., 1000 V), the electrodes are about 9 cm apart, so the voltage gradient across a particle will be small. The sample tends to be diluted during the separation, so theoretically aggregation is not usually a problem, and there is no upper limit to the volume of material that can be processed during a single continuous operation.

The reader is referred to a number of standard texts that explain the construction of CFE machines and the theoretical behavior of biological particles *(1–3)*: It is, however, necessary to state briefly the mode of operation of the machine.

From: *Methods in Molecular Biology, Vol. 19: Biomembrane Protocols: I. Isolation and Analysis*
Edited by: J. M. Graham and J. A. Higgins  Copyright ©1993 Humana Press Inc., Totowa, NJ

## 1.2. The Equipment

The separation chamber (approx $60 \times 10$ cm) consists of two glass plates separated by about 0.7 mm. The separation (chamber) buffer flows vertically down the chamber and separates at the bottom into 90 individually collected outlets, which pass through a peristaltic pump: This pump controls the rate of buffer flow. Two inlet ports are commonly available for applying the sample into the buffer stream. In most applications, the one that is about 2 cm from the cathode is used (*see* Fig. 1). An electric field is applied across the width of the chamber. The electrodes are not in direct contact with the separation buffer; they are immersed in an electrode buffer that circulates from a reservoir. Between the separation buffer and the electrode buffer are two strips of ion exchange resin.

Particles entering the buffer flow experience two forces, a vertical one (buffer flow) and a horizontal one (the field); charged particles will therefore follow a diagonal path down the chamber. Biological particles, being negatively charged under normal pH conditions, will be deflected toward the anode to an extent that depends on their surface charge density at constant buffer flow.

Generally speaking, CFE is most useful for resolving mixed populations of cells *(4–6)*, but it is rather unsuccessful at resolving the majority of subcellular particles in their native state. The technique is very effective at resolving inside-out and right-side-out vesicles from the human erythrocyte *(7)*; it has been applied to the subfractionation of neuraminidase-treated platelet membranes *(8)* for the isolation of lysosomes *(9)*, for the separation of endosomes from trypsin-treated microsomes *(10)*, and for the separation of different Golgi compartments *(11)*.

## 1.3. Experimental Strategy

In a homogenate, apparently only the lysosomes have a surface charge density that under normal conditions, is sufficiently higher than that of all the other subcellular particles that they can be separated by CFE. Even in this case, however, effective resolution of the lysosomes cannot be achieved without extensive washing *(9)* or mild trypsinization of the partially purified subcellular fraction *(10)*. Whether the effect of the washing is to remove contaminating material or whether it elutes adsorbed molecules from the lysosome surface, which might otherwise screen the surface charge, is not clear.

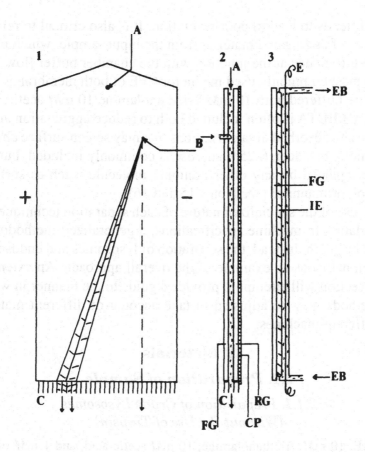

Fig. 1. Diagrammatic representation of continuous-flow electrophoresis apparatus. (1) Front view shows the deflection of particles by the electric field in the downward buffer flow. The dotted line shows the path taken by an uncharged particle. (2) Vertical section through middle of chamber. (3) Vertical section through edge of chamber. **A:** Separation buffer flow from upper reservoir. **B:** Inlet port for sample. **C:** Collection device and peristaltic pump. **CP:** Cooling plate. **E:** Electrode. **EB:** Electrode buffer. **FG:** Front glass plate. **IE:** Ion exchange strip. **RG:** Rear glass plate.

Of the other subcellular particles that comigrate, only the endosome fraction can be successfully resolved by pretreatment with trypsin *(10).* Like the lysosomes, trypsin treatment increases the electronegativity of endosomes, but the concentration of protease needed to do this is higher *(10).*

Even though the process of CFE tends to dilute the sample rather than concentrate it, material at greater concentrations than 1 mg protein/mL tends to lead to poor resolution. It is also critical to remove all traces of aggregated material from the input sample, which might lead to turbulence at the junction with the chamber buffer flow.

The most commonly used media for CFE of both membranes (and cells) are buffered with 10 m$M$ Triethanolamine/10 m$M$ acetic acid, and 1 m$M$ EDTA is often included both to reduce aggregation and to remove any adsorbed divalent cations that may screen surface charge. As osmotic balancer, 0.25$M$ sucrose is commonly included, but this can be replaced by any other nonionic molecule, such as sorbitol, glycerol, or mannitol (*see* Notes 15 and 16)

Because of the specialized nature of each separation technique and the variation in machine specifications, a generalized methodology cannot be given. Instead, the separation of lysosomes and endosomes is given below as an example of the overall approach. An extensive Notes section will attempt to provide a guide to the manner in which the methods may be adjusted to take account of different material and different machines.

## 2. Materials

### 2.1. Preparation of Sample

#### 2.1.1. Preparation of Crude Lysosomes (Without the Use of Trypsin)

1. TAE: 10 m$M$ Triethanolamine, 10 m$M$ acetic acid, and 1 m$M$ EDTA adjusted to pH 7.4 with the base or the acid.
2. Separation medium (SM): 0.25$M$ sucrose in TAE.
3. 50-mL Centrifuge tubes for suitable swing-out rotor of refrigerated low-speed centrifuge.
4. 10-mL Conical centrifuge tubes for suitable swing-out rotor of low-speed centrifuge.
5. 40- to 50-mL Centrifuge tubes for a suitable fixed-angle rotor for a high-speed centrifuge.
6. Dounce homogenizer.

#### 2.1.2. Preparation of Lysosomes and Endosomes (Using Trypsin)

In addition to items listed in Section 2.1.1:

1. 35- to 40-mL Tubes for a suitable swing-out rotor of an ultracentrifuge.
2. 2$M$ (55% [w/w]) Sucrose in TAE.

3. *N*-tosyl-L-phenylalaninechloromethylketone (TPCK)-trypsin (Worthington Biochemical Corp., Freehold, NJ) in SM (0.1 mg/mL).
4. Soya-bean trypsin inhibitor (Sigma Chemical Co., St. Louis, MO) in SM (0.5 mg/mL).
5. Refractometer.

## 2.2. Electrophoresis

1. 6 L of SM (*see* Section 2.1.1.).
2. Electrode buffer: 2 L of 100 m$M$ triethanolamine, 100 m$M$ acetic acid, adjusted to pH 7.4 with base or acid.
3. 100 mL of 1% Bovine serum albumin (BSA) in SM.
4. Continuous-flow electrophoresis apparatus (various models made by Bender and Hobein, Munich, Germany).

# 3. Methods

Important note: The CFE machine (*see* Section 3.2.) needs to be set up about 16 h ahead of the experiment. All g-forces are g$_{av}$.

## 3.1. Preparation of Sample

Carry out all operations at 0–4°C, except where indicated.

### 3.1.1. Separation of Lysosomes (Without Trypsin); Adapted from the Method of Harms et al. (9)

1. Produce a homogenate of the chosen tissue or cultured cells in SM (*see* Note 1).
2. Remove the nuclei and any intact cells by centrifugation at 1000$g$ for 5 min, and repeat this centrifugation on the supernatant.
3. Centrifuge the postnuclear supernatant at 20,000$g$ for 10 min.
4. Discard the supernatant, and resuspend the pellet in SM using four to five gentle strokes of the pestle of a Dounce homogenizer.
5. Repeat steps 3 and 4 another three times (*see* Note 2), and suspend the final pellet in about 10 mL of SM.
6. Centrifuge this suspension three times at 1000$g$ for 5 min to remove any aggregated material (*see* Note 3).
7. Adjust the volume of the supernatant so that the protein concentration is no more than 1.2 mg/mL.

### 3.1.2. Separation of Lysosomes and Endosomes (Using Trypsin); Adapted from the Method of Marsh et al. (10)

1. Prepare a postnuclear supernatant from a homogenate as described in steps 1 and 2 of Section 3.1.1.

2. Using approx 40-mL tubes for a suitable ultracentrifuge swing-out rotor, layer the supernatant onto 0.5 mL of the 2*M* sucrose solution, and centrifuge at 100,000*g* for 30 min (*see* Note 4).
3. Aspirate all the supernatant to just above the band of material; add about 2 mL of Triethanolamine/EDTA buffer; mix by gentle swirling, and decant into a Dounce homogenizer, taking care not to disturb any pelleted material.
4. Homogenize the suspension (four to five gentle strokes of the pestle); adjust to 0.25*M* sucrose using a refractometer (refractive index = 1.3452) and 1 mg protein/mL.
5. Add TPCK-trypsin so that the ratio of trypsin:membrane protein is 0.2–4% by weight (*see* Note 5), and incubate at 37°C for 5 min (*see* Note 6).
6. Shift to 0°C, and then add a fivefold excess of soya-bean trypsin inhibitor.
7. Centrifuge the suspension repeatedly at 1000*g* for 5 min until no further pellet is visible.

### 3.2. Electrophoresis

1. Assemble the separation chamber in accordance with the manufacturer's recommendations (*see* Note 7).
2. Fill the electrode buffer reservoir with electrode buffer, and circulate briefly to fill all channels.
3. Fill the separation chamber with 1% BSA (in SM), and leave overnight (*see* Notes 8 and 9).
4. Wash out the chamber with approx 3 L of SM.
5. Refill the chamber with SM, establish continuity with SM in the upper buffer reservoir, and start the electrode buffer circulation.
6. Adjust the peristaltic pump to produce a chamber buffer flow rate of approx 350 mL/h (*see* Notes 10–12).
7. Set the cooling to 6°C, and when the temperature has reached approx 10°C, start to increase the applied voltage to 1100–1750 V (*see* Notes 10–12) in steps of approx 200 V so that the temperature stays in the range 6–10°C.
8. When the temperature and electrical field are steady, start to inject the sample into the chamber through the inlet port closest to the cathode at 1–4 mL/h (*see* Notes 10–12).
9. Collect and analyze the fractions for appropriate markers (*see* Note 13).
10. At the end of the run while the SM is still flowing, turn down the applied voltage, and reduce the cooling synchronously, so that the temperature rises only slowly to ambient.
11. Wash the chamber with at least 10 L of distilled water (*see* Note 14).

## 4. Notes

1. With cultured cells, aim for approx $2 \times 10^7$ cells/mL. If you are using a tissue, such as rat liver, use approx 100 mL/liver. The method of homogenization will vary with the tissue or cell type (*see* Chapters 3 and 9). If the cells require the use of hypotonic conditions to effect disruption, add a quarter of the volume of SM (at 5× concentration) to the homogenate.
2. The volume of SM used in these washing steps should not be less than half of the original volume of homogenate. Failure to achieve an adequate separation in the CFE may be a result of inadequate washing.
3. The final centrifugation should contain no visible pellet.
4. Using a cushion of dense sucrose prevents the excessive aggregation of particles (mitochondria, lysosomes, and so forth) that sediment at lower relative centrifugal forces (rcf). Alternatively, centrifuge the postnuclear supernatant at 20,000$g$ for 10 min, then centrifuge the supernatant from this, at 100,000$g$ for 30 min, and finally resuspend and combine the two pellets.
5. The amount of TPCK-trypsin required may depend on the source of the material; the amount needed to shift the lysosomes toward the anode is about 0.2%, and that to shift the lysosomes and endosomes is at least 1%.
6. To shift any plasma membrane material toward the cathode (i.e., to reduce electrophoretic migration), the sample could alternatively be incubated with neuraminidase. Only plasma membrane vesicles should have their sialic acid residues exposed to the medium. Menashi et al. *(8)* used this technique with intact platelets, and a similar approach could be tried on intact cultured cells prior to homogenization or on the membrane fraction prior to electrophoresis.
7. Take great care not to mix up the anodal and cathodal ion exchange strips before assembly, and wash them well in distilled water to remove the preserving solution. Always fill up the chamber with distilled water first to check for leaks. It is easier to correct them at this stage than later on. The most common points of leakage are between the bottom of the chamber and the collection manifold, and at the edges of the chamber as a result of faulty alignment of the gasket between the ion exchange strip and the chamber wall.
8. Fill the upper buffer reservoir with SM, and establish fluid continuity with the top of the chamber. This guards against the level of liquid in the chamber falling and ensures that the subsequent flushing out of the BSA solution is efficient.

9. During this step, the glass walls of the chamber become coated with BSA: This reduces loss of subsequent resolution by electroendosmosis *(2)*.

10. The range of permissible buffer flow rates and the maximum permitted current are machine dependent. Since electrical heating is proportional to current, the maximum permissible current is related to the refrigeration capacity of the machine. At the low ionic strength of the buffers employed in membrane fractionation, the maximum current will probably not be reached even at the highest voltage.

11. The magnitude of the applied field obviously controls the horizontal migration of the particles, and generally speaking, the chosen voltage is the one that produces the widest lateral separation between the most and least electronegative material. The combination of buffer flow and applied field should not permit particles to reach either of the electrode strips at any point within the chamber.

12. The lower the buffer flow, the greater the time of residency of the particles in the chamber, and hence the greater will be their deflection by the electric field. Resolution may be helped by the greater angular spread of the deflected band that occurs. However, at too low flow rates, diffusion of particles will reduce resolution. Also the applied voltage tends to be rather less stable. Moreover, the buffer flow rate also limits the rate of which the sample can be injected into the chamber. If the sample flow rate is not reduced to match the chamber buffer flow rate, the stream becomes turbulent and diffuse. High resolution critically depends on a very narrow sample bandwidth at the point of entry.

13. With tissue culture cells, the endosomes and lysosomes are the most electrophoretically mobile species, eluting in fractions 12–20. The Golgi is less electronegative, emerging as two populations around fractions 20 and 22, whereas the bulk of the other membranes (mitochondria, plasma membrane, and so forth) predominates in fractions 24–28. *See* Marsh et al. *(10)* for further details.

14. Unless the CFE machine is going to be used on a routine daily basis, it should be cleaned and disassembled after each run. The chamber will need recoating with albumin before every run.

15. The development of an acceptable CFE method for membranes other than those dealt with here has to be achieved almost empirically. The media that are used for CFE separation of membranes are normally very similar to those used for their routine suspension, e.g., $0.25M$ sucrose, buffered with 10 m$M$ triethanolamine/acetic acid. To obtain adequate electrophoretic separations of different membrane particles, it may be necessary to modulate their surface charge density either by ionic and/or pH manipulation of the medium or chemical/enzymic treat-

ment of the membrane suspension. In this way, because the molecular origins of the surface negative charge on different membrane subfractions are also different, it may be possible to influence differentially this charge and, hence, their electrophoretic mobilities.

16. Variations in the chamber buffer composition that may produce a differential modulation are low concentrations of either EDTA (or EGTA), divalent cations (1–2 m*M*), or alterations to the pH. Although membranes do not tolerate significant shifts in pH, reduction to 6.5 or increase to 8.5 seems not to affect many membrane functions (or at least if it does, the effects are reversible). Moreover, a shift of a single pH unit can have a significant effect on mobility.

# References

1. Hannig, K. and Heidrich, H-G. (1974) The use of continuous preparative free-flow electrophoresis for dissociating cell fractions and isolation of membranous components. *Methods Enzymol.* **31,** 746–761.
2. Hannig, K. and Heidrich, H-G. (1977) Continuous free-flow electrophoresis and its application in biology, in *Cell Separation Methods, Part IV: Electrophoretic Methods* (Bloemendal, H., ed.), Elsevier/North Holland, Amsterdam, pp. 93–116.
3. Graham, J. M., Wilson, R. B. J., and Patel, K. (1986) Free-flow electrophoresis: its application to the separation of cells and cell membranes, in *Cells, Membranes and Diseases* (Reid, E., Cook, G. M. W., and Luzio, J. P., eds.), Plenum, New York, pp. 143–152.
4. Heidrich, H-G. and Drew, M. (1977) Homogenous cell populations from rabbit kidney cortex. Proximal, distal tubule and renin-active cells isolated by free-flow electrophoresis. *J. Cell Biol.* **74,** 780–788.
5. Pretlow, T. G. and Pretlow, T. P. (1979) Cell electrophoresis. *Int. Rev. Cytol.* **61,** 85–128.
6. Wilson, R. B. J. and Graham J. M. (1986) Isolation of platelets from human blood by free-flow electrophoresis. *Clin. Chim. Acta* **159,** 211–217.
7. Heidrich, H-G. and Leutner, G. (1974) Two types of vesicle from the erythrocyte ghost membrane differing in surface charge: separation and characterization by preparative free-flow electrophoresis. *Eur. J. Biochem.* **41,** 37–43.
8. Menashi, S., Weintraub, H., and Crawford, N. (1981) Characterization of human platelet surface and intracellular membranes isolated by free-flow electrophoresis. *J. Biol. Chem.* **256,** 4095–4101.
9. Harms, E., Kern, H., and Schneider, J. A. (1980) Human lysosomes can be purified from diploid skin fibroblasts by free-flow electrophoresis. *Proc. Natl. Acad. Sci. USA* **77,** 6139–6143.
10. Marsh, M., Schmid, S., Kern, H., Harms, E., Male, P., Mellman, I., and Helenius, A. (1987) Rapid analytical and preparative isolation of functional endosomes by free-flow electrophoresis. *J. Cell Biol.* **104,** 875–886.
11. Morré, D. J., Morré, D. M., and Heidrich, H-G. (1983) Subfractionation of rat liver Golgi apparatus by free-flow electrophoresis. *Eur. J. Cell Biol.* **31,** 263–274.

ment of the membrane suspension. In this way, because the molecular origins of the surface negative charge on different membrane subfractions are also different, it may be possible to influence differentially this charge and, hence, their electrophoretic mobilities.

10. Variations in the chamber buffer composition that may produce a differential modulation are low concentrations of either EDTA (or EGTA) divalent cations (1–2 mM), or alterations to the pH. Although membranes do not tolerate significant shifts in pH, reduction to 6.5 or increase to 7.5 seems not to affect many membrane fractions (or, at least if it does, the effect are reversible). Moreover, a shift of a single pH unit can have a significant effect on mobility.

## References

1. Hannig, K. and Heidrich, H.-G. (1974). The use of continuous preparative free-flow electrophoresis for dissociating cell fractions and isolation of membranous components. *Methods Enzymol.* 31, 746–761.

2. Hannig, K. and Heidrich, H.-G. (1977). Continuous free-flow electrophoresis and its application in biology. In *Electrofocusing and Isotachophoresis* (B. J. Radola and D. Graesslin, eds.), Elsevier/North Holland, Amsterdam, pp. 83–116.

3. Heidrich, H.-G., Wilson, R. J., and Platsoucas, C. (1980). Free-flow electrophoresis: its application in the separation of cells and membranes. In *Cell Membranes and Disease* (L. Bolis, G. M. Wei, and L. Leaf, eds.), Raven Press, New York, pp. 143–150.

4. Heidrich, H.-G. and Dew, M. (1977). Homogeneous cell populations from rat kidney cortex: proximal, distal tubule, and renin-active cells isolated by free-flow electrophoresis. *J. Cell Biol.* 74, 780–788.

5. Thedore, T. C. and Heidrich, T. G. (1979). Cell electrophoresis. *Ann. Rev. Cytol.* 61, 85–128.

6. Wilson, R. L. and Graham, J. M. (1984). Isolation of platelets from human blood by free-flow electrophoresis. *Clin. Chim. Acta* 139, C11–C11.

7. Heidrich, H.-G. and Leutner, G. (1974). Two types of vesicles from the erythrocyte ghost membrane differing in surface charge: separation and characterization by preparative free-flow electrophoresis. *Eur. J. Biochem.* 41, 37–43.

8. Monahan, S., Weintraub, H., and Crawford, N. (1981). Characterization of human platelet surface and intracellular membranes isolated by free-flow electrophoresis. *Biochim. Biophys. Acta* 356, 3095–3101.

9. Harris, E., Kern, H., and Schneider, F. A. (1980). Human keratinocytes can be purified from diploid skin fibroblasts by free-flow electrophoresis. *Proc. Natl. Acad. Sci. USA* 77, 6139–6143.

10. Marsh, M., Schmid, S., Kern, H., Harms, E., Male, P., Mellman, I., and Helenius, A. (1987). Rapid analytical and preparative isolation of functional endosomes by free-flow electrophoresis. *J. Cell Biol.* 104, 875–886.

11. Morré, D. J., Morré, D. M., and Heidrich, H.-G. (1983). Subfractionation of rat liver Golgi apparatus by free-flow electrophoresis. *Eur. J. Cell Biol.* 31, 263–274.

CHAPTER 5

# Fractionation of a Microsomal Fraction from Rat Liver or Cultured Cells

## John M. Graham

## 1. Introduction

The membrane organelles of an organized tissue, such as rat liver, that will always form closed vesicles, irrespective of the mode of homogenization, are the tubular membranes of the smooth and rough endoplasmic reticulum. Almost without exception, these vesicles retain their normal orientation, i.e., the cytoplasmic face is outermost. This is particularly clear in the case of the rough endoplasmic reticulum, for the ribosomes remain attached on the outside face. The microsomal (vesicular) fraction of a homogenate will also contain vesicles derived from any domain of the plasma membrane that, in vivo, is not stabilized by structures, such as tight junctions, desmosomes, or an extensive cytoskeleton. Thus, the blood sinusoidal membrane of hepatocytes and the basolateral membranes of enterocytes both tend to form vesicles, whereas hepatocyte contiguous membranes and enterocyte microvillar membranes will remain intact, as long as the homogenization is relatively mild (*see* Chapter 6).

The more severe the homogenization conditions, the greater the tendency for the plasma membrane to vesiculate. This is also true of any extensive Golgi stacks (as in hepatocytes); they will tend to fragment as the severity of the disruptive forces increases. In nitrogen cavitation (*see* Chapter 9), all membrane organelles other than those of the nuclei, mitochondria, and lysosomes tend to form vesicles. Aside from those vesicles that are formed during the process of homogeni-

From: *Methods in Molecular Biology, Vol. 19: Biomembrane Protocols: I. Isolation and Analysis* Edited by: J. M. Graham and J. A. Higgins Copyright ©1993 Humana Press Inc., Totowa, NJ

zation, there are also those that are normally present, which are involved in the processes of membrane synthesis and membrane trafficking, in endocytosis (endosomes), and in secretion. Additionally, there will be vesicles derived from any other membranous organelles that have broken down during the processing of the tissue or cells.

The size of vesicles normally varies from 0.05–0.5 μm, but can be as large as 3.0 μm. Their density ranges from 1.06–1.22 g/cm³ (in sucrose). The density of the rough vesicles is of course raised by the presence of ribosomes, and there will be a broad spectrum of densities attributable to the variable number of ribosomes per vesicle. Smooth vesicles derived from the Golgi tend to be relatively low density (1.07–1.15 g/cm³); in rat liver, the low density is accentuated by the presence of lipoprotein particles within the vesicles. Others may have their densities raised by the association of nonmembrane molecules, e.g., coated or secretory vesicles (*see* Chapter 8).

Total microsomal fractions are easily prepared from a homogenate, since once the mitochondria, lysosomes, and so forth have been removed by pelleting (20,000g for 20 min), the only particles remaining are the microsomes, which are sedimented routinely at 100,000g for 30–60 min. The separation of smooth and rough microsomes can be readily achieved on a sucrose barrier in the presence of CsCl (*1*), but the resolution of all the subfractions of smooth microsomes is quite difficult. There may be size differences between them, but in a sucrose gradient, the hyperosmolarity will reduce or abolish these differences as the vesicles shrink because of water loss.

Other chapters deal specifically with the fractionation of endocytic vesicles (Chapter 7), membrane vesicles from tissue culture cells (Chapter 9), and coated vesicles (Chapter 8). This chapter will be confined to describing:

1. The separation of smooth and rough microsomes;
2. Fractionation of microsomes on top-loaded shallow sucrose gradients; and
3. The use of sucrose gradients followed by low-osmolarity metrizamide or Percoll™ gradients.

A variety of methodologies are given, and the separations that are described (with the exception of the separation of smooth and rough microsomes) should be regarded as examples only. The metrizamide and Percoll gradient separations in particular can be used for diverse samples containing vesicular material; they may be applicable directly to the operator's material, or they may require modification. These subfractionations of smooth microsomal vesicles are used principally in an analytical mode to follow the subcellular processing of macromolecules during synthesis and transport, and of endocytosed ligands; they are not designed to separate a specific population of vesicles.

## 2. Materials

### 2.1. Separation of Smooth and Rough Microsomes from Rat Liver

1. Homogenate produced under standard conditions (*see* Chapter 3, Sections 2.1. and 3.1.).
2. Approximately 50-mL tubes for low-speed swing-out rotor.
3. Approximately 40-mL and 10- to 15-mL polycarbonate tubes for fixed-angle rotors in high-speed and ultracentrifuges, respectively.
4. 0.6$M$ (19% [w/w]) Sucrose and 1.3$M$ (38% [w/w]) sucrose: Both solutions contain 15 m$M$ CsCl and 5 m$M$ HEPES-NaOH (pH 7.6).
5. 0.25$M$ Sucrose in 5 m$M$ HEPES-NaOH (pH 7.6).
6. Syringe and metal cannula (0.5 mm id).

### 2.2. Fractionation of Membrane Vesicles in Top-Loaded Shallow Sucrose Gradients

1. Suitable smooth vesicle fraction prepared from a light-mitochondrial or microsomal pellet (*see* Chapter 3 and Note 1).
2. 0.45$M$ (15% [w/w]), 0.6$M$ (19% [w/w]), 0.73$M$ (23% [w/w]), and 0.87$M$ (27% [w/w]) Sucrose and 0.25$M$ sucrose; all solutions in 5 m$M$ Tris-HCl, pH 8.0.
3. Polycarbonate tubes (approx 5 mL) for suitable high-speed swing-out rotor.
4. Tubes for fixed-angle ultracentrifuge rotor as Section 2.1.
5. Gradient unloader.

### 2.3. Fractionation of Membrane Vesicles in Gradients of Sucrose and Metrizamide

1. A light mitochondrial fraction (20,000$g$ for 10–20 min) or a microsomal fraction isolated from the postnuclear supernatant of a standard rat liver homogenate (*see* Chapter 3 and Note 1).
2. 0.25$M$, 0.77$M$ (24% [w/w]), 1.10 $M$ (33% [w/w]), 1.20$M$ (36% [w/w]), 1.33$M$ (39% [w/w]), and 2.2$M$ (59% [w/w]) Sucrose solutions in 10 m$M$ Tris-HCl, pH 8.0.
3. 10, 15, 20, 25, and 30% (w/v) metrizamide in 10 m$M$ Tris-HCl, pH 8.0.

4. Dounce homogenizer (approx 5-mL capacity).
5. Polycarbonate tubes (12–17 mL) for a suitable ultracentrifuge swing-out rotor and 10- to 15-mL tubes for a fixed-angle rotor.
6. Syringe and metal cannula (0.5 mm id).
7. Refractometer.
8. Gradient unloader (upward displacement preferred).

## *2.4. Fractionation of Membrane Vesicles in a Self-Generated Percoll Gradient*

1. A postnuclear supernatant, postmitochondrial supernatant, or any other membrane sample in 0.25*M* sucrose, 1 m*M* EDTA, and a suitable organic buffer, pH 7.4 (*see* Note 1).
2. Stock Percoll suspension made 0.25*M* with respect to sucrose.
3. 2.5*M* (65% [w/w]) Sucrose in the same buffer as in step 1.
4. Tubes (17–38 mL) for a low-angle fixed-angle or vertical rotor capable of 20,000*g*. Sealed tubes should be used in a vertical rotor (*see* Note 2).
5. Gradient unloader: Use a bottom-puncture apparatus for sealed tubes or any type for fixed-angle tubes.

## 3. Methods

## *3.1. Separation of Smooth and Rough Microsomes from Rat Liver (see Note 3)*

Carry out all operations at 0–4°C. All g-forces are $g_{av}$.

1. Centrifuge the homogenate at 1000*g* for 5 min, and remove the supernatant by decantation.
2. Centrifuge the supernatant at 20,000*g* for 10 min, and again decant the supernatant (*see* Note 4).
3. In 10- to 15-mL polycarbonate tubes, layer 1.5 mL of the 0.6*M* sucrose on top of 3.0 mL of the 1.3*M* sucrose. Then fill up the tubes with the supernatant from step 2.
4. Centrifuge the tubes in a suitable fixed-angle rotor at 100,000*g* for 90 min.
5. Remove the tubes carefully from the rotor at the end of the centrifugation. There are two major fractions: a band (that is often double) at the 0.6/1.3*M* sucrose interface that contains the smooth membranes and a pellet that contains the rough microsomes. Remove the interfacial band using a syringe and metal cannula (*see* Note 5).
6. Discard the remaining supernatant above the pellet, and wash the walls of the centrifuge tube with 0.25*M* sucrose introduced from a syringe while tube is inverted, taking great care not to disturb the pellet.
7. Dilute the smooth microsomes with 2 vol of 5 m*M* HEPES-NaOH (pH 7.6) buffer, and centrifuge at 100,000*g* for 1 h.

8. Resuspend both rough and smooth microsomal pellets in 0.25$M$ sucrose or other suitable solution.

### 3.2. Fractionation of Membrane Vesicles in Top-Loaded Shallow Sucrose Gradients (see Note 6)

1. Make discontinuous gradients from 1 mL each of the 0.45, 0.6, 0.73, and 0.87$M$ sucrose solutions (in 5-mL tubes), and leave at 4°C for at least 6 h to allow the gradient to become linear (*see* Note 7).
2. Suspend the membrane fraction in 0.25$M$ sucrose: Use a syringe for volumes of approx 1 mL (*see* Note 8).
3. Layer 0.5 mL on top of each gradient, and centrifuge at 20,000$g$ for 25 min.
4. Collect in 0.5-mL fractions, and analyze as required (*see* Note 9).

### 3.3. Fractionation of Membrane Vesicles in Gradients of Sucrose and Metrizamide (see Note 10)

1. Resuspend the membrane pellet in 10 mL of 2.2$M$ sucrose so that the final concentration is 1.55$M$. Check the density of the suspension with a refractometer; the refractive index should be approx 1.408. Adjust the concentration of sucrose if necessary.
2. Set up the following discontinuous gradients: 4 mL sample, 2.5 mL of 1.33$M$, 2 mL each of 1.20 and 1.10$M$, and 1 mL each of 0.77 and 0.25$M$ sucrose (*see* Notes 11 and 13), and centrifuge at 100,000$g$ for 1 h.
3. Remove the top two bands (excluding the material floating on the top of the gradient) using a syringe and cannula.
4. Combine the two fractions, dilute with 2 vol of buffer, and sediment at 100,000$g$ for 20 min.
5. Crudely resuspend the pelleted material in about 3 mL of 30% metrizamide, and transfer to a Dounce homogenizer.
6. Homogenize using 10 up-and-down strokes of the pestle (*see* Note 12).
7. Set up the following discontinuous gradients: 3.0 mL sample and 2.5 mL of each of the other metrizamide solutions (*see* Note 11), and centrifuge at 100,000$g$ for 16 h.
8. Unload the gradients into 1- to 1.5-mL fractions by upward displacement.
9. Analyze the fractions for suitable markers (*see* Note 14).

### 3.4. Fractionation of Membrane Vesicles in a Self-Generated Percoll Gradient (see Note 15)

1. Mix the Percoll, membrane suspension, and suspending buffer, if necessary, to give a final Percoll concentration of 27% (density, 1.066 g/cm$^3$).

2. In an appropriate centrifuge tube, layer the suspension over 2.0 mL of 2.5*M* sucrose; centrifuge at 20,000*g* for 2 h.
3. Collect the fractions by bottom puncture or upward displacement, and assay as required (*see* Note 16).

## 4. Notes

1. The source and volume of the material have been specified in each case to show the diversity of these parameters that can be handled and to be able to indicate the separations achieved. All the methods, however, can be applied to any suitable material of any volume: The volumes of gradients and so forth will need the appropriate scaling up or down, and alternative tubes and rotors will need to be chosen. The starting material can be a supernatant or a resuspended pellet from a differential centrifugation scheme. Only in the top-loaded sucrose gradient method (Sections 2.2. and 3.2.), where the volume of sample needed for the highest resolution is small, might it be inconvenient to use a supernatant fraction. Most of the methods have been worked out for rat liver, except the Percoll separations (Section 3.4.), but they all should be applicable to material from any source.
2. Linear self-generating gradients are best produced in the rotors indicated; swing-out rotors and high-angle fixed-angle rotors will produce Percoll gradients that are very "flat" in the middle and steep at the ends.
3. Method adapted from Bergstrand and Dallner *(1)*.
4. Theoretically, the previous centrifugation at 1000*g* is not necessary; the whole homogenate could be centrifuged at 20,000*g*. This, however, tends to result in the loss of a lot of microsomal material into the pellet.
5. Try to aspirate as little of the 1.3*M* sucrose as possible, since this often contains some unpelleted rough microsomes.
6. Method adapted from Grant et al. *(2)*.
7. This particular method has been adapted for a small-volume sample and gradient. If it is scaled up, keep the ratio of sample-to-gradient volume approx 0.1. If you make up the linear gradient using a two-chamber gradient maker, use one with a three- or six-tube manifold. Because of the small volume of the gradient, it is impractical to generate them one at a time using a gradient former.
8. For larger volumes, use a small Dounce homogenizer.
9. Shallow sucrose gradients have been used to separate vesicles containing ligands, vesicles containing receptors, and Golgi vesicles *(2)*.
10. Method adapted from Graham and Winterbourne *(3)*.
11. The discontinuous gradient can be produced by overlayering or underlayering techniques. The latter is more convenient, since the bulk of the gradient can be made up while the sample is being prepared.

12. A Potter-Elvehjem homogenizer can be used as an alternative.

13. Both the sucrose and metrizamide gradients can be adapted to any tube size of a swing-out rotor providing it can generate the requisite rcf. The metrizamide gradients could also be adapted to a vertical rotor with the consequent reduction in run times.

14. Sucrose plus metrizamide gradients have been used to separate Golgi vesicles containing galactosyl transferase, Golgi vesicles containing the sulfotransferase enzyme, and "early" and "late" vesicles involved in the secretion of heparan sulfate *(3)*.

15. Method adapted from Galloway et al. *(4)*.

16. Percoll gradients have been used to separate endosomes from lysosomes and Golgi membranes primarily from tissue culture fibroblasts, but there is no reason why the method cannot be applied to similar material from any source *(4,5)*.

## References

1. Bergstrand, A. and Dallner, G. (1969) Isolation of rough and smooth microsomes from rat liver by means of a commercially available centrifuge. *Anal. Biochem.* **29,** 351–356.

2. Grant, D., Siddiqui, S., and Graham, J. M. (1987) Receptor-mediated endocytosis of enterokinase by rat liver. Preliminary characterization of low-density endosomes. *Biochim. Biophys. Acta* **930,** 346–358.

3. Graham, J. M. and Winterbourne, D. J. (1988) Subcellular localization of the sulphation reaction of heparan sulphate synthesis and transport of the proteoglycan to the cell surface in rat liver. *Biochem. J.* **252,** 437–445.

4. Galloway, C. J., Dean, G. E., Marsh, M., Rudnick, G., and Mellman I. (1983) Acidification of macrophage and fibroblast endocytic vesicles in vitro. *Proc. Natl. Acad. Sci. USA* **80,** 3334–3338.

5. Marsh, M., Schmid, S., Kern, H., Harms, E., Male, P., Mellman, I., and Helenius, A. (1987) Rapid analytical and preparative isolation of functional endosomes by free flow electrophoresis. *J. Cell Biol.* **104,** 875–886.

CHAPTER 6

# Isolation of Plasma Membrane Sheets and Plasma Membrane Domains from Rat Liver

*Laura Scott, Michael J. Schell, and Ann L. Hubbard*

## 1. Introduction

The plasma membrane (PM) of polarized epithelial cells is composed of two physically continuous, but compositionally and functionally distinct domains: basolateral and apical. In hepatocytes, the basolateral domain includes the sinusoidal front, which faces the space of Disse and is involved in exchange of metabolites with the circulation, and the lateral surface, which is contiguous to neighboring hepatocytes. The apical or bile canalicular domain is specialized for the transport of bile acids and is separated from the basolateral domain by tight junctions, which prevent lateral diffusion of membrane proteins and outer leaflet lipids. Biochemical analysis of hepatocyte PM constituents requires the isolation of substantial amounts of this organelle.

In this chapter, we present the materials and methods for the isolation of a PM sheet preparation that is enriched 20- to 40-fold in PM markers *(1)*. This preparation contains substantial amounts of both domains in continuity with each other in ratios approaching those in intact cells. The yield is 10–20%, and the major contaminant is endoplasmic reticulum. PM sheets are suitable for a variety of analytical and preparative uses. For example, we often treat preparative amounts of PM sheets with detergents and purify PM proteins from the extracts *(2)*.

From: *Methods in Molecular Biology, Vol. 19: Biomembrane Protocols: I. Isolation and Analysis*
Edited by: J. M. Graham and J. A. Higgins Copyright ©1993 Humana Press Inc., Totowa, NJ

The PM sheet fraction can also be used as starting material for separation of the two domains. The second part of this chapter describes how to vesiculate PM sheets by gentle sonication and then separate vesicles derived from the different domains by density centrifugation on linear sucrose gradients *(3)*. The separation is most likely owing to the fact that, following cell disruption, elements of the cytoskeleton remain preferentially associated with the cytosolic face of the basolateral domain, making basolaterally derived vesicles more dense. Since basolateral and apical vesicles partially overlap on the gradient, the domain separation is chiefly an analytical procedure, useful for determining the domain location of PM proteins. For example, we have used the domain gradient to study the biogenesis of hepatocyte PM proteins in conjunction with pulse/chase labeling and immunoprecipitation *(4)*.

## 2. Materials

### 2.1. PM Sheets

1. Male Sprague-Dawley rats (125–150 g) (Charles River Breeding Laboratories, Inc., Wilmington, MA). Larger rats give PMs of lesser purity and lower yield because of increased amounts of connective tissue and PM-associated filaments in their livers. For analytical work, one rat provides sufficient material. For preparative work, use up to 40 g of liver at a time (five to six rats).
2. Cheesecloth, Grade 60 (American Fiber and Finishing, Inc., Burlington, MA).
3. A 40-mL Dounce homogenizer with two pestles (Wheaton size A, 0.001- to 0.003-in nominal clearance, and Wheaton size B, 0.0025- to 0.035-in nominal clearance) and 7 mL Dounce with type B (loose) pestle.
4. 50-mL Conical tubes.
5. A low-speed centrifuge, such as the Beckman TJ-6 (Beckman Instruments, Fullerton, CA).
6. An Abbe refractometer.
7. An ultracentrifuge capable of producing an rcf of 113,000$g$ and a swinging bucket rotor of large capacity (e.g., the Beckman SW28).
8. Clear (e.g., polycarbonate) tubes (25 × 89 mm for SW28).
9. Plastic transfer pipets.
10. Stock solutions of protease inhibitors: 0.2$M$ phenylmethylsulfonyl fluoride (PMSF) in 100% ethanol (store at 4°C); 1 mg/mL each of antipain (stable for 1 mo) and leupeptin (stable for 6 mo) in 10% dimethyl sulfoxide (store at –20°C); and aprotinin, 1 mg/mL (store at –20°C).
11. 0.9% (w/v) NaCl can be stored at 4°C for 2 wk.

12. 200 mL of 0.25$M$ STM: 0.25$M$ sucrose (Schwartz/Mann Div., Becton Dickinson & Co., Orangeburg, NY), 10 m$M$ Tris-HCl, 1 m$M$ MgCl$_2$, pH 7.4, density 1.03 g/cm$^3$ ($\eta$ = 1.3450). *See* Section 3.1., step 1 for preparation.
13. 100 mL of 2.0$M$ STM: 2.0$M$ sucrose, 10 m$M$ Tris-HCl, 1 m$M$ MgCl$_2$, pH 7.4, density 1.26 g/cm$^3$ ($\eta$ = 1.4297). *See* Section 3.1., step 1 for preparation.
14. 100 mL of 0.25$M$ sucrose, density 1.03 g/cm$^3$ ($\eta$ = 1.3450). *See* Section 3.1., step 1 for preparation.

## 2.2. PM Domains

1. PM sheets (*see* Section 2.1. and 3.1.).
2. Materials listed in Section 2.1. (steps 6–10).
3. A sonicator (a water-bath-type, 80-W output, available from Laboratory Supplies Co., Hicksville, NY, is suitable).
4. A gradient former/collector, such as the Haake-Buchler Auto Densi-Flow II (Buchler Instruments Inc, Fort Lee, NJ)
5. 7-mL Dounce homogenizer with type B (loose) pestle.
6. 10 mL of 0.25$M$ sucrose, density 1.03 g/cm$^3$ ($\eta$ = 1.3450). *See* Section 3.1., step 1 for preparation.
7. 50 mL of 0.46$M$ sucrose, 10 m$M$ Tris-HCl, pH 7.5, density 1.06 /cm$^3$ ($\eta$ = 1.3557). *See* Section 3.1., step 1 for preparation.
8. 50 mL of 1.42$M$ sucrose, 10 m$M$ Tris-HCl, pH 7.5, density 1.18 /cm$^3$ ($\eta$ = 1.4016). *See* Section 3.1., step 1 for preparation.
9. Large-capacity ultracentrifuge swing-out rotor (e.g., Beckman SW28) and suitable clear (polycarbonate or Ultraclear™) tubes.

## 3. Methods
## 3.1. PM Sheets

1. Prepare all sucrose solutions 24–48 h before use, adjust their pHs, and determine their densities at room temperature with an Abbe refractometer. Filter all solutions through 0.22-μm nitrocellulose, except for the 2.0$M$ STM, which is filtered through 1.2-μm nitrocellulose. Store solutions at 4°C. Alternatively, these solutions can be prepared in advance and stored at –20°C for 1 yr. Check their densities and pHs before use. The volumes of the sucrose solutions given in the methods are for one 6- to 8-g liver.
2. Starve rats for 18–24 h.
3. Precool all glassware and tubes.
4. To analyze the distribution and recoveries at each step of the PM preparation, record the volumes of each fraction and save aliquots. Resuspend unused pellets in a convenient volume of 0.25$M$ STM for analysis (*see* Table 1).

Table 1
Suggested Dilutions for Analysis of PM Prep Fractions

| Fractions | Expected volume, mL, per liver, 6–8 g | Dilutions for 5'-nucleotidase assay | Dilutions for BCA protein assay |
|---|---|---|---|
| H | 40 | 1:50; 1:100 | 1:50; 1:100 |
| S1 | 50 | 1:25; 1:50 | 1:25; 1:50 |
| P1 | 50 | 1:50; 1:100 | 1:20; 1:40 |
| S2 | 40 | 1:20; 1:40 | 1:25; 1:50 |
| P2 | 90 | 1:10; 1:20 | 1:5; 1:10 |
| I | 45 | 1:10; 1:20 | No dilution; 1:2 |
| 1.18 g/m³ | 75 | 1:2; 1:4 | 1:2; 1:4 |
| P3 | 12 | 1:5; 1:10 | 1:50; 1:100 |
| S3 | 40 | 1:2; 1:4 | No dilution |
| PM | 2–3 | 1:50; 1:100 | 1:5; 1:10 |

5. Anesthetize the rat with ether, and kill by decapitation. Drain blood under cold running water, and excise the liver intact.

6. Place liver on a paper towel moistened with ice-cold 0.9% NaCl, and perfuse with the same solution via portal vein until the liver is blanched. Avoid introducing air bubbles into the liver. Quickly weigh the liver, place it in a beaker on ice, and carry out all subsequent procedures at 4°C.

7. Add 4.5 times the liver weight (mL/g) of cold 0.25$M$ STM plus protease inhibitors. (Final concentrations are: 1 m$M$ PMSF, 2 µg/mL each of antipain and leupeptin, and 2 µg/mL of aprotinin). If the liver is bigger than 8 g, divide in half and homogenize separately. Mince the liver with scissors until the pieces are approx 0.5 cm³, and pour the suspension into a 40-mL Dounce homogenizer.

8. Homogenize gently with 10 strokes of the loose-fitting pestle; avoid the formation of bubbles (*see* Note 1). If connective tissue gets stuck on the head of the pestle, wipe it off.

9. Filter the homogenate into a graduated cylinder through four layers of cheesecloth premoistened with 0.25$M$ STM. Rinse the cheesecloth with a small amount of 0.25$M$ STM. Bring the homogenate volume to five times the liver weight (i.e., 20% [w/v]); cover, and invert the cylinder to mix homogenate.

10. During the low-speed centrifugations, the sample volume should be between 25 and 30 mL/tube. Centrifuge the homogenate in 50-mL conical tubes for 5 min at 280$g$ (1100 rpm in Beckman TJ-6). Carefully pour off the supernatant and save it, leaving about 1/3 total volume in the tube to avoid disturbing the loose pellet. Any pellet transferred at this step results in PMs more contaminated with nuclei (*see* Note 2).

11. Resuspend the pellet in 1/2 initial homogenate volume of 0.25*M* STM in a Dounce homogenizer with three strokes of a loose-fitting pestle.

12. Repeat step 10, pooling the supernatants from both centrifugations (S1). If assaying all fractions for recoveries, resuspend the pellet (P1) for analysis.

13. Centrifuge the pooled supernatants 10 min at 1500*g* (2600 rpm in Beckman TJ-6). Pour off the supernatant (S2); the pellet (P2) resulting from this spin is firm and compact.

14. Before resuspending P2, consider that its final volume should be about two times the initial homogenate volume and should also be a multiple of the ultracentrifuge tube volume, less 4 mL/tube. To do this, first resuspend the pellet in one-third the final volume of 0.25*M* STM in a Dounce homogenizer with three strokes of a loose-fitting pestle followed by one stroke with a tight-fitting pestle. Add 2/3 final volume of 2.0*M* STM, and mix the solution thoroughly by gently inverting the cylinder several times. Check the density, and add either 0.25 or 2.0*M* STM to reach a density of 1.18 g/cm$^3$ ($\eta$ = 1.4016).

15. Fill each centrifuge tube to the same volume with the resuspended P2 (approx 32 mL for SW28 tubes), and carefully overlay them with 3 mL of 0.25*M* sucrose.

16. Centrifuge for 1 h at 113,000*g* (25,000 rpm in a Beckman L5-65) at 4°C. Use one-half maximum acceleration, and switch off the brake or use slow brake.

17. Collect the pellicule (I) at the interface with one pass of a plastic transfer pipet. Resuspend to 0.8–1 times the initial homogenate volume in 0.25*M* sucrose. Homogenize gently in a Dounce with three strokes of the loose-fitting pestle. The refractive index should now be <1.3500 (density <1.0460 g/cm$^3$); dilute further if necessary. If assaying all fractions for recoveries, pour off and save the load fraction (1.18 g/cm$^3$), and resuspend the pellet (P3) in the volume listed in Table 1.

18. Fill 50-mL conical centrifuge tubes with the pellicule (I) to a volume of 25–30 mL/tube, and centrifuge for 10 min at 1500*g* (2600 rpm in Beckman TJ-6). Draw off the supernatant (S3) carefully with a pipet; this pellet will be loose. Resuspend the pellet(s) (PM sheets) in 2–3 mL of 0.25*M* sucrose/liver in the 7 mL Dounce homogenizer.

19. Estimate the approximate protein concentration by measuring the absorbance at 280 nm of a 1:20 dilution of plasma membranes in the presence of 1% SDS. Dividing this absorbance by 2.4 gives the protein concentration in mg/mL. Store plasma membranes at –70°C, usually in 1.0-mL aliquots of 2–4 mg/mL.

20. Analyze the distribution and recovery of a plasma membrane marker (e.g., 5'-nucleotidase; *see* Chapter 1) to determine the purity and yield

Table 2
Analysis of Plasma Membrane Fractions[a]

| | Percent Distribution ± SD | | |
| Step | 5'-Nucleotidase | Protein | Relative enrichment[b] ± SD |
|---|---|---|---|
| 1. 280*g* 5 min | | | |
|    Supernate (S1) | 88 ± 7 | 94 ± 3 | <u>0.85 ± 0.12</u> |
|    Pellet (P1) | 12 ± 7 | 6 ± 3 | |
| 2. 1500*g* 10 min | | | |
|    Supernate (S2) | 52 ± 12 | 68 ± 7 | |
|    <u>Pellet</u> (P2) | 48 ± 11 | 32 ± 6 | <u>1.0 ± 0.22</u> |
| 3. Flotation | | | |
|    <u>Interface</u> (I) | 48 ± 15 | 1.9 ± 0.6 | <u>22 ± 6</u> |
|    1.18 g/cm$^3$ layer | 25 ± 5 | 26 ± 3 | |
|    Pellet (P3) | 25 ± 17 | 72 ± 4 | |
| 4. 1500*g* 10 min | | | |
|    Supernate (S3) | 19 ± 6 | 51 ± 8 | |
|    <u>Pellet</u> (PM) | 81 ± 6 | 49 ± 8 | <u>36 ± 21</u> |

[a]Reproduced from ref. *1* by copyright permission of the Rockefeller University Press.
Five separate experiments were analyzed and the results tabulated. The 5'-nucleotidase activity
and protein present in each fraction were measured and their percent distributions within
each step calculated, after correction for recovery. The underlined fraction in each step constituted the starting material for the subsequent step and was normalized to 100%.

[b]Relative enrichment is defined as the ratio of specific 5'-nucleotidase activity in a fraction to the specific activity in the homogenate with the latter normalized to 1. Only relative
enrichments for underlined fractions are presented.

of the preparation. Table 2 gives the expected recovery and distribution
of 5'-nucleotidase throughout the PM preparation. Enrichments can be
calculated after the fractions have been assayed for protein. Table 1
gives dilutions suitable for the bicinchoninic acid (BCA) assay for protein (*see* Chapter 18). Table 3 shows the final enrichments and recoveries of several enzyme markers in the PMs (*see* Note 3).

21. Examine the PM sheet preparation using phase microscopy before proceeding further. Apply a sufficient volume to the slide (approx 20 µL)
so that the PM sheets can be seen moving between the slide and the
cover slip; use a 25× objective. Long, dense strips that look like "Ys"
should be prominent in the preparation, and no large, round spheres
(nuclei) should be present. (*See* Notes 1 and 2 and Fig. 1a.)

Table 3
Recoveries and Enrichments of Various Organellar Markers
During the Isolation of Hepatocyte Plasma Membrane Sheets[a]

| Marker | Recovery,[b] % | Enrichment,[b] -fold |
|---|---|---|
| Protein | 0.4 ± 0.13 | — |
| 5'-Nucleotidase (PM) | 17.4 ± 6 | 39 ± 10 |
| Alkaline phosphodiesterase I (PM) | 17.0 ± 5.6 | 40 ± 9 |
| Asialoglycoprotein binding activity (PM) | 16.3 ± 6 | 47 ± 18 |
| NADH-cytochrome c reductase (ER) | 0.3 ± 0.2 | 1 ± 0.5 |
| Glucose-6-phosphatase (ER) | 0.5 | 0.8 |
| β-N-acetylglucosaminidase (lysosomes) | 0.22 ± 0.13 | 0.66 ± 0.3 |
| Cytochrome oxidase (mitochondria) | 0.12 ± 0.07 | 0.22 ± 0.1 |
| DNA (nuclei) | 0.26 | 0.3 |
| Galactosyltransferase (Golgi) | 1.0 | 1.9 |

[a]Reproduced from ref. *1* by copyright permission of the Rockefeller University Press.
[b]Reported as mean ± SD ($n \geq 3$), when available.

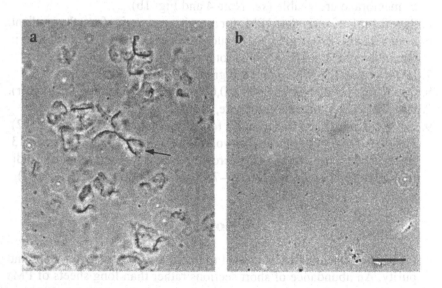

Fig. 1. Phase contrast microscopy of plasma membrane sheets and vesicles. (**a**) The arrow points to plasma membranes that are visible as long interconnected black lines with membrane attached to their ends and sides. (**b**) When vesiculation is complete, no strips of lateral surface are present, and very few distinct pieces of membrane are visible. Very short pieces of membrane, which appear as black dots, may remain visible. Bar = 20 μm.

### 3.2. PM Domains

1. Prepare the sucrose solutions as described in Section 3.1., step 1. The volumes of the sucrose solutions given in the methods will make $2 \times 32$ mL gradients.
2. Dilute fresh or frozen PMs to 1 mg/mL in $0.25M$ sucrose and protease inhibitors (*see* Section 3.1., step 7). If using SW28 tubes, use 4 mg for each gradient.
3. Sonicate 2 mL aliquots in 14-mL polystyrene tubes using a bath sonicator containing an ice slurry. Sonicate for bursts of 10 s with 1-min pauses on ice between bursts. Depending on the strength of the sonicator, between 3 and 20 bursts will be needed to complete vesiculation. Vesiculation can also be achieved using a Polytron homogenizer (Brinkman, Westbury, NY) fitted with a 12-mm tip at setting 8 (*see* Note 4).
4. Every one to five bursts, monitor the vesiculation by light microscopy using phase contrast optics at 25×. Each tube should be monitored independently. Vesiculation is complete when no, or very few, dark strips of membrane are visible (*see* Note 4 and Fig. 1b).
5. Save 0.1 mL of vesiculated PMs for assaying recoveries from the gradient.
6. Pour a 32-mL linear gradient of sucrose from 0.46 to $1.42M$ sucrose in clear tubes for Beckman SW28 rotor (or equivalent).
7. Layer vesicles (3.9 mL) on each gradient with a transfer pipet.
8. Centrifuge 16–20 h at 72,000$g$ (20,000 rpm in a Beckman SW28 rotor). Use 1/2 acceleration and no brake, or brake to 800 rpm.
9. Collect 3 mL fractions from the top of the gradient (fractions 1–12). Resuspend the pellet to a final volume of 3 mL to make fraction 13. Invert each fraction to mix the sucrose and measure their refractive indices. Store fraction as aliquots at –70°C (and analyze; *see* Notes 4–7).

## 4. Notes

1. Proper homogenization is crucial to obtaining PMs of good yield and purity. An abundance of short sections rather than long sheets of PMs may be caused by homogenizing too vigorously in step 8 (*see* Fig. 1a).
2. If PMs are contaminated with nuclei, which appear under the microscope as 10-μm spheres, pour less supernatant off of the pellet in step 10.
3. In recent fractionations, as little as 50% of a PM enzyme activity (e.g., alkaline phosphodiesterase I) initially present in P2 has been recovered in the three subsequent fractions. Some of this activity subsequently reappears in the PM fraction, which causes an apparent overrecovery

from the interface (I) fraction. The reasons for these discrepancies are not currently clear.

4. Incomplete vesiculation, resulting from undersonication, is the most commonly encountered problem in the domain separation. It is characterized by >25% of the PM markers being found in the pellet and apical proteins having a basolateral-like distribution. To correct, increase the number of sonication bursts until no membrane strips are visible (*see* Fig. 1b). Oversonication is a much less likely problem. It may result in the basolateral markers having a more apparently apical distribution in the gradient.

5. The distributions of basolateral and apical membrane markers in the gradient can be assessed by immunoblotting or enzyme assay. To Western blot with domain-specific antibodies, use 0.1–0.5 mL of each gradient fraction (*see* Chapter 20). To assay for the distribution of the apical marker 5'-nucleotidase, use 50 µL of each gradient fraction (*see* Chapter 1). Both $K^+$-stimulated *p*-nitrophenylphosphatase activity and ouabain-sensitive $Na^+/K^+$-ATPase activity have a basolateral distribution on the domain gradient (*see* Note 6). Figure 2 shows a typical domain distribution as monitored by immunoblotting for a basolateral and an apical PM protein. The peak of apical PM is found in fractions 5–6 (1.10 g/cm$^3$), and the peak of basolateral PM is found in fractions 8–9 (1.14 g/cm$^3$).

6. To assay the $K^+$-stimulated *p*-nitrophenylphosphatase *(7)*, dilute 50 µL of each PM fraction with 1 mL of either Buffer I (11.5 m$M$ KCl, 3.45 m$M$ MgCl$_2$, 5.75 m$M$ theophylline, 1.15 m$M$ EDTA, and 50.0 m$M$ Tris-HCl, pH 7.4) or Buffer II (11.5 m$M$ NaCl, 3.45 m$M$ MgCl$_2$, 5.75 m$M$ theophylline, 1.15 m$M$ EDTA, and 50.0 m$M$ Tris-HCl, pH 7.4). Start the reaction by adding 100 µL of substrate mix (34.5 m$M$ *p*-nitrophenylphosphate, 0.0575% (w/v) saponin, 50.0 m$M$ Tris-HCl, pH 7.4). Incubate at 37°C for 60 min. Stop the reaction by the addition of 100 µL of 10$M$ NaOH. Measure absorbance at 405 nm. The millimolar extinction coefficient of the reaction product is 18.0. Subtract the amount of $K^+$-independent phosphatase activity (Buffer II) from the phosphatase activity assayed in the presence of $K^+$ (Buffer I). To determine rigorously the distribution of the ouabain-sensitive $Na^+/K^+$-ATPase activity, assay the domain fractions with Buffer I ±5.75 m$M$ ouabain as above (*see* Chapter 1).

7. The majority of the vesicles produced will be oriented with their ectoplasmic domain on the outside of the vesicles. This was determined elsewhere *(3)*, where $^{125}$I-wheat germ agglutinin was bound to the basolateral PM surface during perfusion at 4°C, followed by PM vesicle preparation. Some 75–90% of the $^{125}$I-wheat germ agglutinin bound to the vesicles was accessible to release by *N*-acetyl-D-glucosamine.

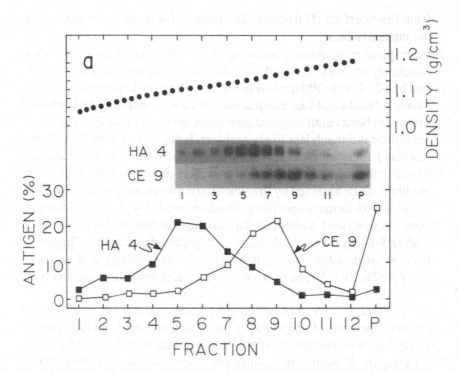

Fig. 2. Distributions of an apical and basolateral marker in the PM domain gradient. The distributions of HA4 (apical) and CE9 (basolateral) in these gradients were determined by immunoblotting and plotted as a percentage of the recovered antigen. Reproduced from ref. *3* by copyright permission of the Rockefeller University Press.

## References

1. Hubbard, A. L., Wall, D. A., and Ma, A. (1983) Isolation of rat hepatocyte plasma membranes. I. Presence of the three major domains. *J. Cell Biol.* **96,** 217–229.
2. Hubbard, A. L., Bartles, J. R., and Braiterman, L. T. (1985) Identification of rat hepatocyte plasma membrane proteins using monoclonal antibodies. *J. Cell Biol.* **100,** 1115–1125.
3. Bartles, J. R., Braiterman, L. T., and Hubbard, A. L. (1985) Endogenous and exogenous domain markers of the rat hepatocyte plasma membrane. *J. Cell Biol.* **100,** 1126–1138.
4. Bartles, J. R., Feracci, H. M., Stieger, B., and Hubbard, A. L. (1987) Biogenesis of the rat hepatocyte plasma membrane in vivo: comparison of the pathways taken by apical and basolateral proteins using subcellular fractionation. *J. Cell Biol.* **105,** 1241–1251.

5. Widnell, C. C. and Unkeless, J. C. (1968) Partial purification of a lipoprotein with 5' nucleotidase activity from membranes of rat liver cells. *Proc. Natl. Acad. Sci. USA* **61**, 1050–1057.
6. Ames, B. N. (1986) Assay of inorganic phosphate, total phosphate, and phosphatases. *Methods Enzymol.* **8**, 115–118.
7. Stieger, B., Marxer, A., and Hauri, H-P. (1986) Isolation of brush-border membranes from rat and rabbit colonocytes: is alkaline phosphatase a marker enzyme? *J. Membr. Biol.* **2**, 19–31.

5. Widnell, C. C. and Unkeless, J. C. (1968). Partial purification of a lipoprotein with 5'-nucleotidase activity from membranes of rat liver cells. Proc. Natl. Acad. Sci. USA 61, 1050-1057.

6. Ames, B. N. (1966) Assay of inorganic phosphate, total phosphate and phosphatases. Methods Enzymol. 8, 115-118.

7. Stieger, B., Marxer, A. and Hauri, H-P. (1986). Isolation of brush-border membranes from rat and rabbit colonocytes. Is alkaline phosphatase a marker enzyme? J. Membr. Biol. 91, 19-31.

CHAPTER 7

# The Analysis of Receptor-Mediated Endocytosis by Centrifugation in Isoosmotic Gradients

*Terry C. Ford*

## 1. Introduction

### 1.1. Background

Receptor-mediated endocytosis is an important mechanism by which the cell is able to take in macromolecules from the external environment and carry them through the necessary metabolic processes. Studies of the sequential steps of these events require methods of following the fate of the molecule from its internalization to its degradation in lysosomes or its transfer to other intracellular compartments.

The results of a number of studies have indicated that, in many cases, the surface receptors involved are grouped in "coated pits" *(1)*. When ligands bind to their receptors, the coated pits invaginate to form deeper wells, which then close at the surface to form vesicles containing the ligands, finally pinching off from the surface and moving into the cell. The coated vesicles thus formed become larger by fusion with other vesicles of the endocytic pathway *(2,3)*, followed by compartmentalization within these larger vesicles, with the result that the components of the coated pits are recycled to the surface membrane while the ligands continue their journey to the site of degradation.

At a later stage in the degradative pathway, the vesicles fuse with primary lysosomes in which the products of degradation are first detected; they then move to, or become, secondary lysosomes from which the exodus of the degradation products from the organelle takes place.

From: *Methods in Molecular Biology, Vol. 19: Biomembrane Protocols: I. Isolation and Analysis*
Edited by: J. M. Graham and J. A. Higgins  Copyright ©1993 Humana Press Inc., Totowa, NJ

The pathways have been investigated by a number of techniques, including centrifugation on density gradients, usually of sucrose.

Membrane-bound vesicles are very sensitive to changes in the osmotic environment, taking in water and swelling in hypoosmotic conditions and losing water and shrinking in hyperosmotic conditions. The changes in volume lead to changes in the buoyant densities of the vesicles, and this, in turn, can result in information gathered from centrifugation studies being misleading with regard to changes occurring during endocytosis. It is therefore important to maintain a stable environment, as close as possible to the in vivo osmolality of the vesicles, in order to obtain a true picture of events.

The methods to be described result from studies of the internalization and degradation of asialo-glycoproteins by suspensions of isolated rat liver hepatocytes, prepared by whole-organ perfusion *(4,5)*. However, the centrifugation techniques described here can be applied to the uptake of any ligand, using intact tissue or isolated cells. The density gradients were prepared using a mixture of Nycodenz™ and sucrose solutions in order to maintain the whole gradients as close as possible to the in vivo osmotic environment of hepatocytes, using a modified procedure from Ford and Rickwood *(6)*. This chapter is concerned primarily with the use and modulation of the gradients to study the endocytic vesicles involved in the degradative pathway. Chapter 8 of this vol. is concerned, specifically, with the isolation and analysis of coated vesicles. Chapters 23 and 24 of *Biomembrane Protocols: II. Architecture and Function* deal with the methods used to study ligand processing.

## 1.2. Experimental Strategy

The methodology described in this chapter is designed to follow the route of a radioisotopically labeled ligand through the cell, from its internalization to the exodus of the degraded products, by looking at changes in the cell-associated ligand with time since internalization. To do this, it is necessary to be able to time the periods of internalization accurately. During incubation at 4°C, ligand can bind to cell-surface receptors, but is not internalized at that temperature *(7)*. Internalization of the ligand is started by incubation at 37°C and stopped by addition of ice-cold medium.

The procedures described in Section 3.2. ensure that the only radiolabeled ligand present is that which has been internalized by the cells during the incubation period at 37°C. Under the conditions described, we can follow the fate of the ligand until the products of its degradation leave the cell. Any such labeled products that have left the cell during the incubation period chosen will therefore have been washed away before homogenization of the cells.

During centrifugation through a density gradient, larger particles, in general, sediment faster than smaller and keep sedimenting until they reach a point in the gradient where the density of the gradient medium is equal to the density of the particle, at which point the particle sediments no further. Using these criteria, we can centrifuge the cytoplasmic extracts on density gradients and, from their sedimentation behavior, deduce something of the size of the labeled vesicles after various periods of incubation. These experiments provide information about the changes in size of the vesicles during their pathway to degradation and give the opportunity to examine vesicles at different stages of development. This is possible because, as described in the protocol to follow, different fractions of the gradient will contain vesicles of endocytosed ligand that are of increasing size and, therefore, of faster sedimentation rates, progressively from the top of the gradient to their isopycnic position in the gradient.

The actual site of degradation of the [125]I-labeled ligand is not indicated, since the labeled products leave the site and the cell too quickly. Although this chapter is intended to deal mainly with the centrifugation techniques required to follow the path of the ligand through the cell, it may be useful to refer to a method of looking at the site of degradation itself. If the ligand is labeled with [125]I-tyramine-cellobiose, the labeled degradation products are trapped at the site of degradation *(8)*, which enables these products to serve as markers for the organelles involved in the degradation process. By these means, it is possible to follow the progress of the ligand through the gradient as before, but by assaying for acid-precipitable radioactivity (nondegraded ligand) and acid-soluble activity (degraded ligand), we can obtain a clearer view of the degradation sites *(see* Note 8).

## 2. Materials

### 2.1. Preparation of the Hepatocyte Suspension (9,10)

1. Preperfusion buffer: 8.0 g NaCl, 0.4 g KCl, 0.06 g $Na_2HPO_4 \cdot 2H_2O$, 0.047 g $KH_2PO_4$, 0.20 g $MgSO_4 \cdot 7H_2O$, 2.05 g $NaHCO_3$, dissolved in water to a final volume of 1000 mL and a pH of 7.1 *(see* Note 1).

2. Perfusion buffer: 35 mg collagenase (130–150 U/mg solid) and 100 μL of 1.0M CaCl$_2$—make up to 50 mL with preperfusion buffer. Gas both these buffers with 95% O$_2$/5% CO$_2$ to pH 7.5 during perfusion.

3. Incubation buffer: 8.5 g NaCl, 0.4 g KCl, 0.06 g Na$_2$HPO$_4$· 2H$_2$O, 0.047 g KH$_2$PO$_4$, 0.20 g MgSO$_4$·7H$_2$O, 4.76 g HEPES, 0.29 g CaCl$_2$·2H$_2$O. Make up to 1000 mL with water, and adjust to pH 7.5 with NaOH. The osmolality of this solution is approx 300 mOsm.

4. Incubation buffer plus 1.0% BSA.

5. Shaking water bath.

6. Low-speed centrifuge, swing-out rotor for 15 and 50 mL tubes, and fixed-angle rotors.

7. Nylon mesh.

8. Perfusion apparatus (*see* Fig. 1).

## 2.2. Binding and Internalization of Ligand

1. Ligand solution: The ligand, in this example $^{125}$I-asialofetuin (7), is dissolved at 2 μg/mL in incubation/BSA buffer.

2. Ice-water bath.

3. 37°C Water bath.

4. 0.25M Sucrose solution containing 4 mM EDTA, 1 mM HEPES-NaOH (pH 7.2).

5. Dounce homogenizer with tight-fitting pestle.

## 2.3. Centrifugation Materials (see Note 2)

1. Nycodenz solution: 40% (w/v) Nycodenz, 1 mM EDTA, 1 mM HEPES NaOH, pH 7.2.

2. Sucrose solution: 0.25M sucrose, 1 mM EDTA, 1 mM HEPES-NaOH, pH 7.2.

3. Low-speed centrifuge, 15- and 50-mL tubes.

4. Centrifuge with swing-out rotor capable of 85,000g.

5. Gradient mixer and gradient collector (upward displacement).

## 2.4. Detection of Acid-Precipitable/ Acid-Soluble Fractions

1. 60% (v/v) Trichloroacetic acid (TCA).

2. 2-mL Sample tubes.

3. Refrigerated bench-top centrifuge.

4. Gamma counter.

5. $^{125}$I-tyramine-cellobiose labeled ligand.

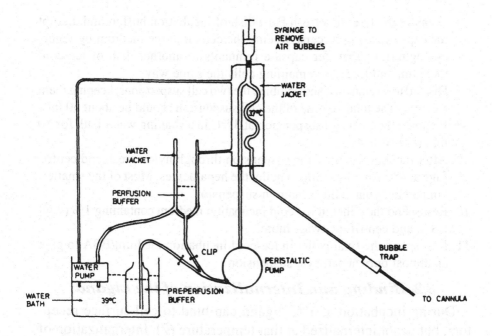

Fig. 1. The apparatus setup for whole organ perfusion shown diagrammatically.

# 3. Methods

## 3.1. Preparing the Hepatocyte Suspension (9,10)

1. Expose the liver of a deeply anesthetized rat, and insert the cannula (Fig. 1) into the vena porta about 1 cm from its junction with the liver with the preperfusion buffer flowing at about 10 mL/min. Once the cannula is secured in place, increase the flow rate to 50 mL/min, and immediately sever the inferior vena cava to permit free flow (*see* Note 3).
2. Now kill the rat by puncturing the diaphragm. Free the liver by cutting away all attachments, taking care not to damage the liver capsule.
3. Continue the preperfusion with the liver *in situ* for about 10 min. Ensure that the lobes lie in their natural position, allowing flow of buffer to all parts.
4. Switch to the perfusion buffer, at the same flowrate, for about 10 min, after which it should be possible to see disintegration of the liver within the capsule.
5. Disconnect the cannula, and transfer liver to a Petri dish containing ice-cold incubation buffer. Remove all extraneous tissue from the liver, taking care not to damage the capsule.

6. Transfer the liver to a fresh Petri dish of incubation buffer, and disrupt the capsule, using forceps, freeing the cells into the medium by vigorous agitation. Transfer capsule remnants to another dish of ice-cold medium, and free any remaining cells the same way.
7. Discard any remnants, and combine the two cell suspensions, keeping them ice-cold. The total volume of the cell suspension should be about 50 mL.
8. Incubate the total cell suspension at 37°C in a shaking water bath for 30 min (*see* Note 4).
9. After incubation, sieve the suspension through nylon mesh and centrifuge at 4°C for 30 s at 50*g* to pellet the hepatocytes. Most of the smaller, nonparenchymal cells remain in suspension.
10. Resuspend the pellet in ice-cold incubation medium containing 1% (w/v) BSA, and centrifuge twice more.
11. Resuspend the final pellet in ice-cold incubation medium/BSA to give an almost pure hepatocyte suspension.

### 3.2. Binding and Internalization of the Ligand

During incubation at 4°C, ligand can bind to cell-surface receptors, but is not internalized at that temperature *(7)*. Internalization of the ligand is started by incubation at 37°C and stopped by addition of ice-cold medium (*see* Note 5).

1. Incubate the hepatocyte suspension at 4°C for 60 min in the presence of $^{125}$I-labeled asialofetuin (ligand solution) to allow ligand to bind the cell-surface receptors (*see* Note 6).
2. Wash the cells three times in the ice-cold incubation/BSA medium to remove unbound ligand. Cell concentration should be $5–10 \times 10^6$/mL after the final pellet is resuspended.
3. Divide the suspension into the number of aliquots required for the various time periods chosen for internalization of the ligand.
4. Add warm incubation medium to each aliquot, and transfer to a 37°C water bath to start the incubation. At the end of the chosen time, add ice-cold medium to stop internalization (*see* Note 5).
5. Wash each aliquot in ice-cold 0.25*M* sucrose, 4 m*M* EDTA, and 1 m*M* HEPES-NaOH (pH 7.2), which releases any externally bound ligand from the surface receptors. Thus, only internalized ligand is associated with the cells.

### 3.3. Centrifugation Analysis

1. Prepare a preformed density gradient (30-mL vol), using the Nycodenz and sucrose solutions (Section 2.3.) and a mechanical gradient mixer. The density range of this gradient is $1.04–1.20$ g/cm$^3$.

2. Homogenize the chopped liver in the sucrose solution in the Dounce homogenizer (10 strokes of the pestle) and sediment the nuclei at 700$g$ for 5 min.
3. Top-load each gradient with 4 mL of the postnuclear supernatant and centrifuge for 45 min at 85,000$g$ at 4°C.
4. After centrifugation, a prominent, light brown/beige colored band is visible toward the bottom of the gradient, and other faint bands may be visible above it.
5 Fractionate the whole gradient into 2-mL fractions by upward displacement, and assay each fraction to determine the distribution of the radiolabel and of various marker enzymes (*see* Notes 6–10).

### 3.4. Detection of Acid-Precipitable/ Acid-Soluble Fractions

The detection of the distribution of acid-soluble and acid-precipitable fractions in the gradient can be carried out using a $^{125}$I-tyramine-cellobiose labeled ligand, which retains the products of degradation at the site of degradation. The preparation is exactly the same as that described earlier, apart from the substitution of ligands.

Having fractionated the gradient as before, assay each fraction for the presence of acid-precipitable/acid-soluble radioactivity.

1. Pipet 0.2- to 0.5-mL aliquots into samples tubes; make up to 1.0 mL with water, and add 0.2 mL 60% TCA.
2. Mix well and keep at 4°C for at least 1 h. Centrifuge at 500$g$ for 15–30 min at 4°C.
3. Separate the pellet and supernatant, and count both in a gamma counter.

Also assay for a lysosomal enzyme marker, e.g., β-*N*-acetylglucosaminidase or β-galactosidase (*see* Chapter 1).

### 4. Notes

1. The preperfusion buffer contains no calcium, and this allows the breakdown of the desmosomal cell-cell junctions, which are calcium dependent. The presence of calcium is required in the perfusion buffer for the reason that the collagenase action in the breakdown of the intercellular matrix is calcium dependent.
2. The gradient solutions and preformed gradients (Sections 2.3. and 3.3.) were used for the experiments described in the Methods section. However, it is possible that some modification could be advantageous under different circumstances. The preparation of Nycodenz gradients that will maintain isoosmolality throughout has been fully described elsewhere (*6,11,12*), but, in brief, 30% (w/v) Nycodenz in water has an

osmolality of 300 mOsm and a density of 1.159 g/cm$^3$. Solutions up to
this concentration can be made isoosmotic with mammalian cells (about
290–320 mOsm) by addition of suitable osmotic balancers, such as NaCl,
sucrose, or glucose. Solutions of Nycodenz at higher concentration than
30% will be hyperosmotic, but to a far less degree than sucrose solution
of the same density *(6,11,12)*. The gradients so prepared will maintain
an isoosmotic profile throughout the gradient.

3. The initial slow flow rate is necessary to allow the cannula to be inserted and secured, before severing the inferior vena cava, without the risk
   of inflating the blood vessels and damaging the liver.

4. During the enzymatic dissociation of the tissue, large numbers of receptor
   molecules are lost from the cell surface, but a large proportion of these
   can be replaced during incubation at 37°C for 30–45 min *(7)*.

5. To ensure accurate timing of the very short incubation periods, it is
   necessary to be able to start and stop the incubation almost instanta-
   neously. This can be achieved by keeping the aliquots on ice and having aliquots of the incubation medium in small flasks in a shaking water
   bath at about 45°C. To start an incubation, add the sample aliquot to
   one of the flasks containing three times the volume of the sample, mix,
   and transfer to a 37°C bath. To stop the incubation, pour the contents of
   the sample flask into another flask, on ice, containing three times the
   volume of the sample and mix quickly.

6. The experiments described have been fully reported *(4,5)*, and the results analyzed. If the incubation periods chosen are 0.5, 1.0, 2.5, 15, and 30 min, the
   results that may be expected are shown diagrammatically (Fig. 2). It can be
   seen that, after 0.5 min incubation, very little ligand has been internalized and is
   in the form of small, slowly sedimenting vesicles. After 1.0 min, the amount of
   internalized ligand is more than doubled, with a peak of slowly sedimenting
   particles, and populations of faster sedimenting particles. After 2.5 min, the
   slowly sedimenting species are still apparent, but much larger populations of
   the faster sedimenting particles have developed. Almost all the internalized
   ligand is incorporated in the faster sedimenting particles after 15 min, and after
   30 min, the much reduced peak of activity indicates that degraded ligand has
   left the cell.

7. If the cytoplasmic extracts from the incubation periods indicated are centrifuged under the same conditions, but with the centrifugation time increased
   from 45 to 180 min, the peak radioactivity for all vesicles is found at a buoyant
   density of close to 1.11 g/cm$^3$, showing that the banding densities found
   after 45 min centrifugation are owing to their sedimentation rates,
   and therefore to differences in their sizes, not differences in the buoyant density of particles after the various incubation periods.

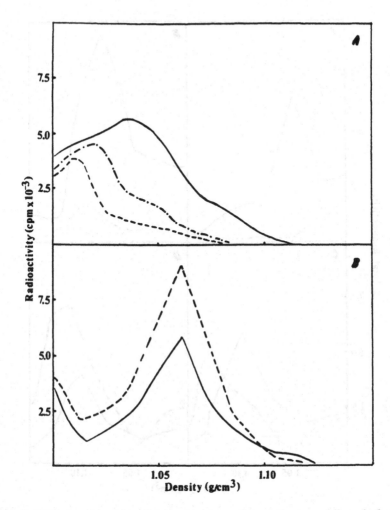

Fig. 2. Differences in the sedimentation rates of the endocytosed ligand after various incubation periods to internalize the ligand. Panel A: after 30 s – – –; after 60 s – · – · – · –; after 150 s ——. Panel B: after 15 min – – –; after 30 min ——. Note the decrease in cell-associated radioactivity after 30 min indicating the exodus of degraded ligand from the site of degradation. Compare with Fig. 3, where the degraded (acid-soluble) activity is retained in the lysosomal area.

8. The results that may be expected from the experimental protocol using the $^{125}$I-tyramine-cellobiose ligand *(5)* are shown diagrammatically (Fig. 3). No soluble radioactivity is detected in the gradient fractions at the shorter incubation periods, but appears first at 15 min incubation at 37°C. The soluble activity coincides with the acid-precipitable activity at a buoyant density of 1.10 g/cm$^3$, that of the large endosomes. The

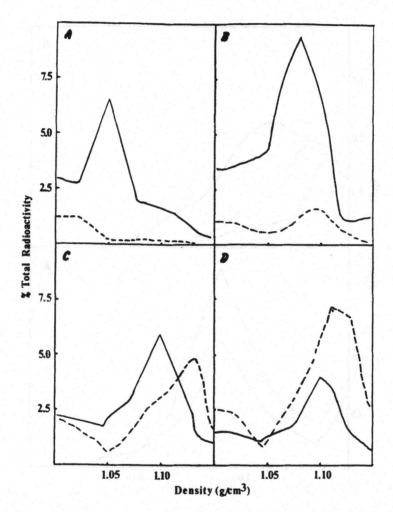

Fig. 3. The distribution of degraded and undegraded ligand. Acid-soluble radioactivity (degraded) – – –; acid-precipitable activity (undegraded) ———; in the Nycodenz/sucrose gradients after incubation with [125]I-tyramine-cellobiose ligand for (**A**) 1 min, (**B**) 15 min, (**C**) 60 min, or (**D**) 90 min.

amount of soluble activity increases with incubation time and acid-precipitable activity decreases. After a 30-min incubation period, an additional peak of soluble activity appears at a density of about 1.13 g/cm³, which is coincident with a well-defined shoulder of a lysosomal marker enzyme, β-*N*-acetylglucosaminidase (not shown), the main peak of lysosomal activity being at a buoyant density of 1.15 g/cm³. After 60- and

90-min incubation periods, the soluble species continues to increase with the precipitable activity decreasing, and an increasing amount of the soluble activity coinciding with the peak of lysosomal activity. There is also a significant shift in the profile of the lysosomal activity between incubation periods of 0 and 90 min, the average buoyant density of the peak decreasing with time.

9. The protocols described here demonstrate the wide range of information regarding the fate of internalized ligands that may be obtained from these simple experiments. They show the early events after internalization and indicate a time-scale for the pathway from internalization to degradation together with information about the sizes of vesicles. By trapping the degradation products within the organelles concerned, we can see at least five steps in the pathway from plasma membrane to a small, slowly sedimenting vesicle to a larger, faster sedimenting vesicle. Two types of lysosomal activity are detected, the first coincident with the larger endosome and later appearing in a denser region of the gradient, coincident with the main peak of lysosomal activity.

10. The inert, nonionic nature of both Nycodenz and sucrose allows examination of the effect of inhibitors without interference from the gradient medium. The advantage of these gradients over those of sucrose alone is owing to the much lower osmolality of the Nycodenz/sucrose gradients, which allows increased resolution of the different species of organelles and the possibility of isolating vesicles at different stages of their progress along the metabolic pathway.

## References

1. Ashwell, G. and Harford, J. (1982) Carbohydrate specific receptors of the liver. *Ann. Rev. Biochem.* **51**, 531–554.
2. Harford, J. and Ashwell, G. (1982) in *The Glycoconjugates*, vol. 4 (Horowitz, M. I., ed.), Academic, New York, pp. 27–55.
3. Harford, J., Wolkoff, A. W., Ashwell, G., and Klausner, R. D. (1983) Monensin inhibits intracellular dissociation of asialoglycoprotein from its receptor. *J. Cell Biol.* **96**, 1824–1828.
4. Kindberg, G. M., Ford, T. C., Rickwood, D., and Berg, T. (1984) Separation of endocytic vesicles in Nycodenz gradients. *Anal. Biochem.* **142**, 455–462.
5. Berg, T., Kindberg, G. M., Ford, T. C., and Blomhoff, R. (1985) Intracellular transport of asialoglycoproteins in rat hepatocytes—evidence for two sub-populations. *Exp. Cell Res.* **161**, 285–296.
6. Ford, T. C. and Rickwood, D. (1982) Formation of isotonic Nycodenz gradients for cell separations. *Anal. Biochem.* **124**, 293–398.
7. Tolleshaug, H. (1981) Binding and internalization of asialoglycoprotein by isolated rat-liver hepatocytes. *Int. J. Biochem.* **13**, 45–51.

8. Pittman, R. C., Green, S. R., Attie, A. D., and Steinberg, D. (1979) A radio-iodinated intracellularly-trapped ligand for determining the sites of plasma protein degradation *in vivo*. *J. Biol. Chem.* **254,** 6876–6879.
9. Seglen, P. O. (1979) Separation approaches for liver and other cell sources, in *Cell Populations, Methodological Surveys (B), Biochemistry,* vol. 9 (Reid, E., ed.), Ellis Horwood, Chichester, England, pp. 25–46.
10. Berg, T. and Blomhoff, R. (1983) Fractionation of mammalian cells, in *Iodinated Density Gradient Media* (Rickwood, D., ed.), IRL at Oxford University Press, Oxford, UK, pp. 173,174.
11. Rickwood, D. (1983) Properties of iodinated density gradient media, in *Iodinated Density Gradient Media* (Rickwood, D., ed.), IRL at Oxford University Press, Oxford, UK, pp. 1–21.
12. Ford, T. C. and Graham, J. M. (1991) Formation of density gradients, in *An Introduction to Centrifugation*, Bios Scientific, Oxford, UK, pp. 39–46.

CHAPTER 8

# The Isolation of Clathrin-Coated Vesicles and Purification of Their Protein Components

## Antony P. Jackson

## 1. Introduction

### 1.1. Background

Coated vesicles are implicated in a number of receptor-mediated transport events within eukaryotic cells. In particular, they are required for the export of newly synthesized lysosomal enzymes from the trans-Golgi network, and at the plasma membrane they are responsible for receptor-mediated endocytosis of selected ligands (1). Coated vesicles are the best characterized of all transport organelles, owing in large measure to the ease with which they can be purified.

The most striking feature of coated vesicles is the outer polyhedral lattice composed of the protein clathrin (Figs. 1 and 2C). The functional unit is the triskelion, each leg of which contains a single molecule of clathrin heavy-chain and either one of two clathrin light-chain isoforms: LCa or LCb. Inherent flexibility at the vertex and elbow allows the triskelion to polymerize into the hexagons and pentagons of the coated vesicle (Fig. 1).

Underneath the clathrin shell lies an inner layer of proteins that, in vitro, promotes the polymerization of triskelions. For this reason they have been called assembly proteins (2). In their full biological role they link triskelions with the cargo of membrane-bound receptors at

From: *Methods in Molecular Biology, Vol. 19: Biomembrane Protocols: I. Isolation and Analysis*
Edited by: J. M. Graham and J. A. Higgins Copyright ©1993 Humana Press Inc., Totowa, NJ

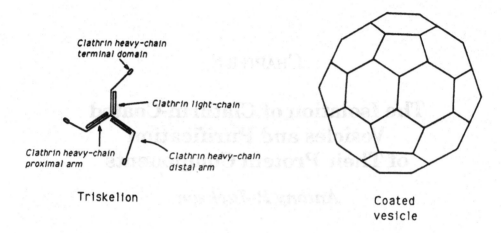

Fig. 1. The clathrin triskelion. In the assembled coated vesicle, a triskelion is centered on every vertex. The terminal domains face inward and bind the adaptors *(1,3)*.

the core of the vesicle, and to emphasize this aspect it has been suggested that they be called adaptors *(3)*. Each adaptor contains proteins with mol wts in the 50-kDa and 20-kDa range associated with two distinct 100-kDa molecules called adaptins. Structural *(4–6)* and sequence analysis *(7–10)* reveals three families of adaptins termed, α, β, and γ. Adaptors from the Golgi are heterodimers of a β-class adaptin and γ adaptin, whereas the plasma-membrane-derived adaptors contain one molecule each of β and α adaptin *(3,6)*. Given their distributions, a plausible supposition is that β adaptins provide a binding site for clathrin, whereas α and γ adaptins may help define specificity by selecting different cargo molecules. A summary of the main structural components of coated vesicles is given in Table 1.

Coated vesicles were first isolated from brain *(11)*, and this remains the most popular choice because the yield is significantly greater than for other tissues. However, note that the components of brain-coated vesicles have a number of unique features including inserted sequences in the clathrin light-chains *(12–14)*, a brain-specific α-class adaptin *(7)*, and at least two brain-specific clathrin-associated proteins *(15,16)*. Brain-coated vesicles also contain tubulin in higher amounts than those from other tissues *(17)*.

## 1.2. General Strategy

There are a number of isolation protocols that work well and are described in the literature. All use a buffer system that stabilizes coated vesicles and require the initial purification of a postmitochondrial supernatant. Very pure preparations can be obtained by density gradient centrifugation in either sucrose or Ficoll/$D_2O$ *(11,18,19)*. However, we prefer a protocol first described by Campbell et al. *(20)* and modified by Bar-Zvi and Branton *(21)*, which has the advantage of simplicity. In this method, coated vesicles are purified by a series of differential centrifugation steps followed by gel filtration.

High concentrations of Tris will dissociate both triskelions and adaptors from the core of the coated vesicles. These components can be separated by gel filtration *(18)*. The Golgi and plasma membrane-derived adaptors can be further purified on hydroxylapatite and are therefore referred to as HAI and HAII respectively *(3–5)*. Clathrin light-chains LCa and LCb show a remarkable heat stability and remain soluble after boiling. This property can be exploited to purify them away from clathrin heavy-chain *(22,23)*. Ion-exchange chromatography is then used to separate the two isoforms.

## 2. Materials

### 2.1. Coated Vesicle Isolation

1. Isolation buffer: 10 m*M* MES-NaOH, pH 6.5, 100 m*M* KCl, 1 m*M* EGTA, 0.5 m*M* $MgCl_2$, 0.2 m*M* dithiothreitol (DTT), 0.02% (w/v) sodium azide, 50 mg/L phenylmethylsulfonyl fluoride (PMSF) (add PMSF to the isolation buffer immediately before use from a stock solution of 50 mg/mL in isopropanol).
2. Isolation buffer plus 12.5% (w/v) Ficol 400 and 12.5% (w/v) sucrose.
3. Waring blender.
4. Refrigerated centrifuge and high capacity rotor (Sorval GS-3 or equivalent).
5. Ultracentrifuge and rotor (Beckman 45 Ti or equivalent).
6. Chromatography column (about 2.6 × 100 cm) packed with Sephacryl S-1000 superfine, fraction collector, pump, and UV monitor.
7. Fine muslin.
8. Dounce homogenizer.

### 2.2. Separation of Triskelions and Adaptors

1. Dissociation buffer: 0.5*M* Tris-HCl, pH 7.0, 1 m*M* EDTA, 0.2 m*M* DTT, 0.02% (w/v) sodium azide, 50 mg/L PMSF (freshly added, *see above*).

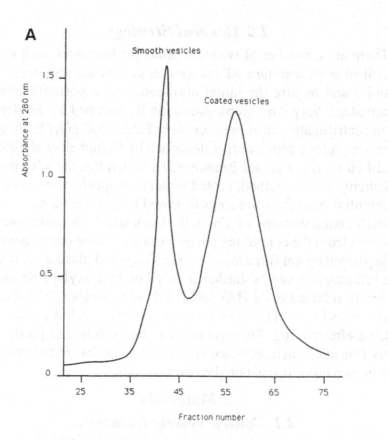

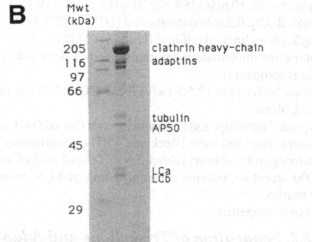

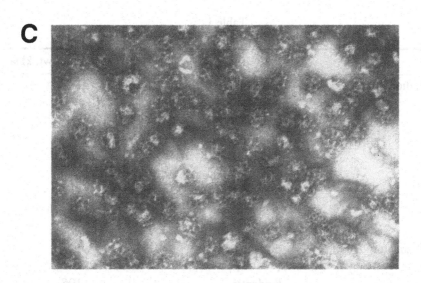

Fig. 2. *(opposite page and above)* (**A**) Purification of coated vesicles by gel filtration on Sephacryl S-1000 Superfine. (**B**) Purified coated vesicles analyzed by SDS-PAGE. (**C**) Electron micrograph of the same coated vesicle preparation shown in **A** and **B**. Negatively stained with uranyl acetate (magnification 40,000×) .

2. Chromatography column (2.6 × 100 cm) packed with Sepharose CL-4B, fraction collector, pump, and UV monitor.
3. Ammonium sulfate (or ultrafiltration device).

## 2.3. Separation of Adaptor Complexes

1. Starting buffer: 10 m$M$ phosphate buffer, pH 8.4, 0.1$M$ NaCl, 0.2 m$M$ DTT, 10% (v/v) glycerol, 50 µg/mL PMSF (added fresh).
2. Limit buffer: 0.5$M$ phosphate buffer, pH 8.4, 0.1$M$ NaCl, 0.2 m$M$ DTT, 10% (v/v) glycerol, 50 µg/mL PMSF (added fresh).
3. About 5 mL hydroxylapatite in a suitable chromatography column, gradient maker, fraction collector, pump, and UV monitor (*see* Note 5).
4. Dialysis tubing.
5. High-speed centrifuge and suitable fixed-angle rotor (e.g., 10 × 12 mL).

## 2.4. Purification of Clathrin Light-Chain Isoforms

1. DEAE buffer: 10 m$M$ Tris-HCl, pH 7.5, 0.2 m$M$ EDTA, 0.2 m$M$ DTT, 0.02% (w/v) sodium azide.
2. DEAE buffer with 0.3$M$ NaCl.
3. About 10 mL DEAE cellulose (Whatman DE-52) in a suitable chromatography column, gradient maker, fraction collector, pump, and UV monitor.

Table 1
The Major Protein Components of Coated Vesicles

|  | Protein | Apparent mol wt, kDa |
|---|---|---|
| Triskelion | clathrin heavy-chain | 180 |
|  | clathrin light-chain LCa (brain-type) | 38 |
|  | clathrin light-chain LCb (brain-type) | 36 |
|  | clathrin light-chain LCa (nonbrain) | 34 |
|  | clathrin light-chain LCb (nonbrain) | 32 |
| HAII adaptor (plasma membrane derived) | $\alpha_A$ adaptin (brain-specific) | 112 |
|  | $\beta$ adaptin | 106 |
|  | $\alpha_C$ adaptin | 105 |
|  | AP50 | 50 |
|  | AP17 | 17 |
| HAI adaptors (Golgi-derived) | $\beta'$ adaptin | 110 |
|  | $\gamma$ adaptin | 104 |
|  | AP47 | 47 |
|  | AP20 | 20 |
|  | $\alpha/\beta$ tubulin (most prominent in brain coated vesicles) | 56/53 |

4. Dialysis tubing.
5. Ultrafiltration device.

# 3. Methods

Carry out all operations at 4–6°C. All g-forces are $g_{av}$.

## 3.1. Coated Vesicle Isolation

1. Obtain the brains as fresh as possible, and store on ice until use (*see* Notes 1 and 2). Remove remnants of the spinal cord, and cut into small cubes a few centimeters square. Homogenize with an equal volume of isolation buffer in a Waring blender (use short bursts of 15–20 s each followed by intervals of about 10 s).

2. Pour the homogenate into 500-mL bottles, and centrifuge at 10,000*g* for 30 min.

3. Decant the clear red supernatant through fine muslin to remove floating fat debris. Discard the thick creamy pellet.

4. Centrifuge the supernatants at 120,000*g* for 30 min.

5. Using a tight-fitting Dounce homogenizer, resuspend the creamy pellet in the minimum volume of isolation buffer (approx 3 mL/pellet).

6. To the resuspended pellets add an equal volume of isolation buffer containing Ficoll/sucrose, and centrifuge for 40 min at 25,000*g*.

7. Pour the slightly cloudy supernatant into a measuring cylinder. Add 3 vol of isolation buffer to dilute the Ficoll/sucrose. Centrifuge at 120,000*g* for 1 h.

8. Discard the supernatant (*see* Note 3). Thoroughly resuspend the glassy pellet in not more than 20 mL of isolation buffer. If necessary use a Sorval Omnimixer at moderate speed. The suspension will have a milky, slightly yellow appearance.

9. Centrifuge for 10 min at 5000*g* in a bench-top centrifuge to remove aggregates. Load onto a 500-mL column of Sephacryl S-1000 superfine pre-equilibrated with isolation buffer. Elute at about 20 mL/h, and collect 5-mL fractions.

10. Two peaks elute from the column, both with a milky, slightly bluish hue. The first peak, eluting at the void volume, consists of smooth vesicles devoid of most proteins. The second, broader peak contains coated vesicles (Figs. 2A–C). Pool the second peak fractions. Centrifuge 120,000*g* for 1 h, and either resuspend in a minimum volume of isolation buffer or, if further purification is required, proceed to Section 3.2.

### *3.2. Separation of Triskelions and Adaptors*

1. Resuspend the coated vesicle pellets from Section 3.1., step 10 in about 3 mL of dissociation buffer. Homogenize and incubate on ice for 1 h.

2. Clarify the suspension by ultracentrifugation at 120,000*g* for 1 h.

3. Rehomogenize the pellet with another 3 mL of dissociation buffer, and repeat step 2.

4. Combine the two supernatants, and load onto a 2.6 × 100 cm column of Sepharose CL-4B pre-equilibrated in dissociation buffer. Elute with dissociation buffer at 15 mL/h, and collect 5-mL fractions.

5. Three peaks should elute from the column. The first, a small peak, elutes at the void volume and contains residual uncoated vesicles. The second peak contains triskelions and the third adaptors (Fig. 3).

6. Separately pool triskelion and adaptor peak fractions, and concentrate either by adding ammonium sulfate to 50% saturation or by ultrafiltration.

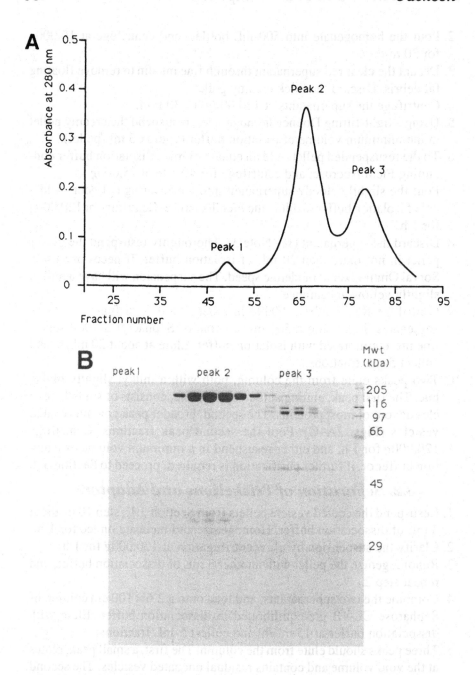

Fig. 3. Gel filtration of triskelions and adaptors on Sepharose CL-4B. (A) Elution profile. (B) Analysis of peak fraction by SDS-PAGE.

7. For long-term storage, freeze both triskelions and adaptors in dissociation buffer at −80°C.

### 3.3. Separation of Adaptor Complexes (see Notes 5–7)

1. Dialyze the adaptors (from Section 3.2., step 6) extensively against at least 2 × 100 vol of starting buffer.
2. Remove any aggregates by centrifugation for 10 min at 10,000g.
3. Load at least 5 mg protein onto the hydroxylapatite column at 20 mL/h, and wash with at least 10 column vol of starting buffer.
4. Elute with a linear gradient composed of 30 mL of starting buffer and 30 mL of limit buffer. Collect 1-mL fractions.
5. The HAI group elutes first at about 0.1M phosphate followed by the HAII group at just over 0.2M phosphate (Fig. 4, *see* Notes 7 and 8).

### 3.4. Purification of Clathrin Light-Chain Isoforms (see Notes 9–12)

1. Concentrate triskelions to about 5 mg protein/mL in dissociation buffer. Immerse about 5 mL of this solution in a boiling water bath for 10 min.
2. Allow to cool, then centrifuge the turbid solution at 100,000g for 1 h.
3. Dialyze the supernatant against 2 × 100 vol of DEAE buffer.
4. Load the dialyzed supernatant onto a 10-mL column of DEAE-cellulose pre-equilibrated in DEAE buffer and wash through with 5–10 column vol of DEAE buffer.
5. Apply a 200-mL linear gradient of 0–0.3M NaCl in DEAE buffer. The first peak corresponding to LCb elutes at about 0.2M NaCl, whereas LCa elutes at about 0.25M NaCl (Fig. 5).
6. Concentrate each pool by ultrafiltration, dialyze back into DEAE buffer, and store at −80°C.

## 4. Notes

1. Either cow or pig brains are convenient sources. We usually process 40 pig brains at a time. This amount, however, requires three separate centrifugations in a Beckman 45 Ti rotor at Section 3.1., step 4 of the protocol. Typically yields are about 40–50 mg coated vesicles/kg brain wet wt. Other tissues yield lower amounts.
2. Although best used fresh, tissue can be frozen at −80°C for up to 2 mo. To thaw, place the frozen brain in a beaker with an equal volume of isolation buffer, and place the beaker in a basin of warm water.

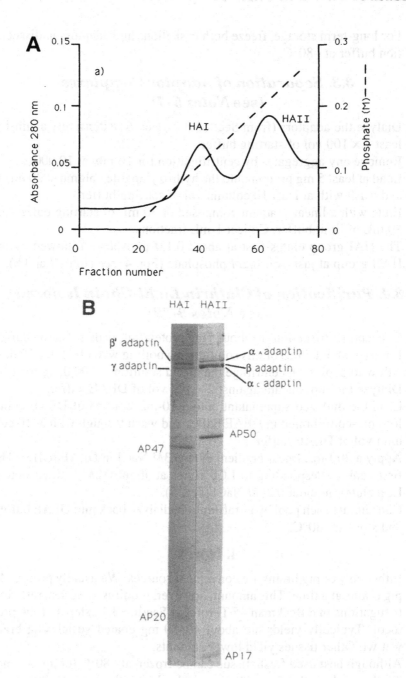

Fig. 4. Separation of HAI and HAII complexes by hydroxylapatite chromatography. (A) Elution profile. (B) Analysis of complexes by SDS-PAGE.

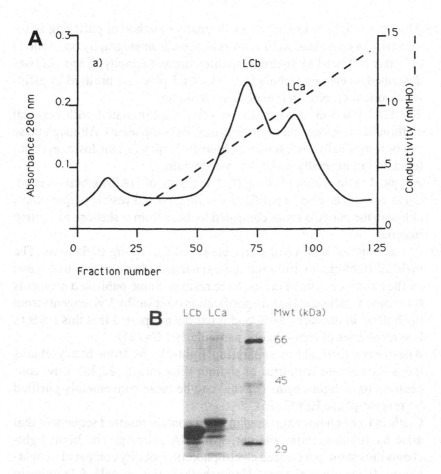

Fig. 5. Purification of clathrin light-chain isoforms by ion-exchange chromatography. (A) Elution profile. (B) Analysis of peak fractions by SDS-PAGE.

3. For many purposes the coated vesicles at Section 3.1., step 9 are already sufficiently pure. If the aim is to purify coated vesicle proteins, the gel-filtration step can be omitted and the pellet further processed as described in Section 3.2.

4. Coated vesicles required for functional studies are best kept at 4°C in isolation buffer with protease inhibitors PMSF (50 µg/mL), leupeptin (10 µg/mL), and aprotinin (10 µg/mL).

5. Adaptins are extremely sensitive to proteolysis, and on occasions we have even found it necessary to autoclave fraction-collector tubes before use. The buffers must always be treated with fresh PMSF.

6. Ahle et al. *(6)* have described an alternative method of purifying adaptors using a combination of ion-exchange chromatography on Mono Q columns followed by hydroxylapatite chromatography. Keen *(24)* has described an elegant technique in which adaptors are purified by affinity chromatography on immobilized triskelions.

7. The HAI fraction in particular is often contaminated with residual clathrin heavy-chain and other high-mol-wt components. Although these can be removed by ion-exchange chromatography *(4)*, the losses encountered do not normally make this worthwhile.

8. A typical yield is about 0.3 mg HAI and 2 mg of HAII/kg brain wet wt.

9. Light-chains can also be purified by boiling coated vesicle preparations, although the purity is lower compared to those from triskelions as starting material.

10. About 1 mg of light-chains are recovered per 10 mg triskelions. The yield of light-chains from the ion-exchange column seems to depend on the oxidation state of the cysteine resides. Some published protocols for coated vesicle and clathrin purification omit sulfhydryl reagents from the buffers at the early steps, and it has been reported that this leads to low recoveries of light-chains, particularly LCa *(25)*.

11. Alternative methods of separating light-chains from heavy-chains include treatment with urea or sodium thiocyanate *(22,26)*. Low concentrations of light-chains (<1 mg) can be more conveniently purified by reverse-phase HPLC *(25)*.

12. Clathrin light-chains expressed in brain contain inserted sequences that arise by tissue-specific alternative mRNA splicing. The brain light-chains thus show a decreased electrophoretic mobility compared to light-chains from other tissues. Nevertheless, the so-called "nonbrain light-chains" occur in brain as minor species and will be present in a typical preparation *(12–14)* (*see* Fig. 5B).

## References

1. Brodsky, F. M. (1988) Living with clathrin: its role in intracellular membrane traffic. *Science* **242,** 1396–1402.
2. Zaremba, S. and Keen, J. H. (1983) Assembly polypeptides from coated vesicles mediate reassembly of unique clathrin coats. *J. Cell Biol.* **97,** 1339–1347.
3. Pearse, B. M. F. (1989) Characterization of coated-vesicle adaptors: their reassembly with clathrin and with recycling receptors. *Meth. Cell Biol.* **31,** 229–243.
4. Pearse, B. M. F. and Robinson, M. S. (1984) Purification and properties of 100-kd proteins from coated vesicles and their reconstitution with clathrin. *EMBO J.* **3,** 1951–1957.
5. Manfredi, J. J. and Bazari, W. L. (1987) Purification and characterization of two distinct complexes of assembly polypeptides from calf brain coated vesicles

that differ in their polypeptide composition and kinase activities. *J. Biol. Chem.* **262**, 12,182–12,188.

6. Ahle, S., Mann, A., Eichelsbacher, U., and Ungewickell, E. (1988) Structural relationships between clathrin assembly proteins from the Golgi and the plasma membrane. *EMBO J.* **7**, 919–929.

7. Robinson, M. S. (1989) Cloning of cDNAs encoding two related 100-kD coated vesicle proteins (α-adaptins). *J. Cell Bio.* **108**, 833–842.

8. Kirchhausen, T., Nathanson, K. L., Matsui, W., Vaisber, A., Chow, E. P., Burne, C., Keen J. H., and Davis, A. E. (1989) Structural and functional division into two domains of the large (100 to 115-kDa) chains of the clathrin-associated protein complex AP-2. *Proc. Natl. Acad. Sci. USA* **86**, 2612–2616.

9. Ponnambalam, S., Robinson, M. S., Jackson, A. P., Peiperl, L., and Parham, P. (1990) Conservation and diversity in families of coated vesicle adaptins. *J. Biol. Chem.* **265**, 4814–4820.

10. Robinson, M. S. (1990) Cloning and expression of γ-adaptin, a component of clathrin-coated vesicles associated with the Golgi apparatus. *J. Cell Biol.* **111**, 2319–2326.

11. Pearse, B. M. F. (1975) Coated vesicles from pig brain: Purification and biochemical characterization. *J. Mol. Biol.* **97**, 93–98.

12. Jackson, A. P., Seow, H-F., Holmes, N., Drickamer, K., and Parham, P. (1987) Clathrin light chains contain brain-specific insertion sequences and a region of homology with intermediate filaments. *Nature* **326**, 154–159.

13. Kirchhausen, T., Scarmato, P., Harrison, S. C., Monroe, J. J., Chow, E. P., Mattaliano, R. J., Ramachandran, K. L., Smart, J. E., Ahn, A. H., and Brosius, J. (1987) Clathrin light-chains LCA and LCB are similar, polymorphic, and share repeated heptad motifs. *Science* **236**, 320–324.

14. Jackson, A. P. and Parham, P. (1988) Structure of human clathrin light-chains: conservation of light-chain polymorphism in three mammalian species. *J. Biol. Chem.* **263**, 16,688–16,695.

15. Ahle, S. and Ungewickell, E. (1986) Purification and properties of a new clathrin assembly protein. *EMBO J.* **5**, 3143–3149.

16. Ahle, S. and Ungewickell, E. (1990) Auxilin, a newly identified clathrin-associated protein in coated vesicles from bovine brain. *J. Cell Biol.* **111**, 19–21.

17. Kelly, W. G., Passaniti, A., Woods, J. W., Daiss, J. L., and Roth, T. F. (1983) Tubulin is a molecular component of coated vesicles. *J. Cell Biol.* **97**, 1191–1199.

18. Keen, J. H., Willingham, M. C., and Pastan, I. H. (1979) Clathrin-coated vesicles: isolation, dissociation and factor-dependent reassociation of clathrin baskets. *Cell* **16**, 303–312.

19. Pearse, B. M. F. (1982) Coated vesicles from human placenta carry ferritin, transferrin and immunoglobulin G. *Proc. Natl. Acad. Sci. USA* **79**, 451–455.

20. Campbell, C., Squicciarini, J., Shia, M., Pilch, P. F., and Fine, R. E. (1984) Identification of protein kinase as an intrinsic component of rat liver coated vesicle. *Biochemistry* **23**, 4420–4426.

21. Bar-Zvi, D. and Branton, D. (1986) Clathrin-coated vesicles contain two protein kinase activities. *J. Biol. Chem.* **261**, 9614–9621.

22. Lisanti, M. P., Shapiro, L. S., Moskowitz, N., Hua, E. L., Puszkin, S., and Schook, W. (1982) Isolation and preliminary characterization of clathrin-associated proteins. *Eur. J. Biochem.* **125,** 463–470.
23. Brodsky, F. M., Holmes, N. J., and Parham, P. (1983) Tropomysin-like properties of clathrin light chains allow a rapid, high-yield purification. *J. Cell Biol.* **96,** 911–914.
24. Keen, J. H. (1987) Clathrin assembly proteins: affinity purification and a model for coat assembly. *J. Cell Biol.* **105,** 1989–1998.
25. Parham, P., Brodsky, F. M., and Drickamer, K. (1989) The occurrence of disulphide bonds in purified clathrin light chains. *Biochem. J.* **257,** 775–781.
26. Ungewickell, E. (1983) Biochemical and immunological studies on clathrin light chains and their binding sites on clathrin triskelions. *EMBO J.* **2,** 1401–1408.

CHAPTER 9

# Isolation of Membranes
# from Tissue Culture Cells

## John M. Graham

## 1. Introduction
### 1.1. Homogenization
#### 1.1.1. Cell Type

The isolation of membranes from cultured mammalian cells poses a number of problems. First and foremost is the problem of homogenization, which is particularly difficult with suspension culture cells, as opposed to those that grow as an adherent monolayer. The choice of subcellular fractionation procedure depends on the results of the homogenization. For a review of the problems associated with the isolation of plasma membranes from tissue culture cells, *see* Graham *(1)*.

By and large, the methods of fractionation have been developed for soft tissues such as rat liver, and they are not easily extrapolated to cultured cells, except perhaps to monolayers of hepatocytes. The efficacy of the homogenization of a coarsely minced soft tissue is achieved by a mixture of mechanical and liquid shear forces. The relative ease of fractionation of the products depends on the existence of structures that stabilize certain areas of the plasma membrane (e.g., desmosomes and tight junctions) and consequently favor cell rupture at specific weak points (*see* Chapter 6).

A confluent cell monolayer that is detached from the substratum by scraping with a rubber policeman comes closest to behaving in the manner of a soft tissue; whereas cells in suspension (e.g., lymphocytes or

From: *Methods in Molecular Biology, Vol. 19: Biomembrane Protocols: I. Isolation and Analysis*
Edited by: J. M. Graham and J. A. Higgins Copyright ©1993 Humana Press Inc., Totowa, NJ

Ehrlich Ascites cells) are difficult to handle, particularly if they are relatively small and if their nuclei occupy most of the cytoplasm.

### 1.1.2. Homogenization Medium

The choice of homogenization medium depends on cell type. The ideal choice is an isoosmotic medium, but many cell lines, particularly those in suspension, are difficult to disrupt without prior swelling of the cells by osmotic stress. Media such as buffered $0.25M$ sucrose and a standard liquid shear homogenizer (e.g., a Dounce) may be applicable to monolayer cells if a satisfactory level of cell disruption (e.g., more than 90%) can be obtained with about 20 strokes of the pestle. If the use of an isoosmotic medium requires 40 or 50 passes of the pestle to effect an acceptable level of cell breakage, then the cells will need to be rendered more fragile by osmotic swelling.

Although hypoosmotically swollen cells may be easier to homogenize, the released organelles (nuclei, mitochondria, lysosomes, and so forth) will also be exposed to a low osmolarity, and they may therefore swell and release some of their contents. DNA released from nuclei is potentially detrimental for it leads to almost irreversible aggregation of subcellular material. Ions that stabilize the nuclei, such as $Mg^{2+}$ and $Ca^{2+}$, also tend to stabilize the plasma membrane against disruption.

Hypoosmotic media such as a 5–10 m$M$ organic buffer (Tris, Tricine, HEPES, and so on) or an inorganic buffer such as 1 m$M$ sodium bicarbonate may be used. Cations can be added to the buffer before or immediately upon homogenization depending on their relative stabilization of the plasma and nuclear membranes. It is also advantageous to return the homogenate to isoosmolarity by the addition of sucrose or other osmotic balancer. The released organelles can also be protected to some extent by the soluble proteins of the cytoplasm: The effect is maximized if the cell:homogenization volume is minimized.

Hypoosmotic media can be avoided if the cells are allowed to take up glycerol from the surrounding medium: The cell contents thus become hyperosmotic. The osmotic gradient across the plasma membrane can be modulated and magnified by transfer of the glycerol-loaded cells to an isoosmotic or only slightly hypoosmotic medium. In this way some cells will burst without using a homogenizer, others will still need such a device.

### 1.1.3. Nitrogen Cavitation

An alternative method for the disruption of cultured cells is nitrogen cavitation. The cells are suspended in a suitable isoosmotic medium and in a stainless steel pressure vessel, the suspension is equilibrated with nitrogen at about 5500 kPa (800 psi). A delivery tube that leads from the stirred suspension exits from the chamber and leads to a needle valve. When the valve is opened the cell suspension is ejected and the sudden drop to atmospheric pressure causes the cells to burst. Cell rupture is caused by the sudden expansion of nitrogen dissolved within the cytosol (bursting balloon effect), and the rapid appearance of bubbles of nitrogen gas in the liquid medium, which causes gaseous shear at the cell surface.

The methods for homogenizing tissue culture cells not only depend on the cell type, but also the time the cells have been in continuous culture, the growth conditions, and the confluency of the cells. It is not possible to give a generally applicable method to all cultured cells. The methods should therefore be taken merely as a guide, which may need modification (*see* Notes).

### 1.2. Subcellular Fractionation

It is impossible to give any definitive method for the fractionation of membranes from tissue culture cells. The size of fragments and/or vesicles from the plasma membrane in particular will vary from cell to cell and with the chosen method of homogenization. The methods given in Section 3.2. provide a strategy to enable the operator to work out an optimal method.

The strategy depends to some extent on the volume of homogenate. If the volume is less than about 30 mL, the entire material could be fractionated initially on a rate zonal gradient in a swing-out rotor in a high-speed centrifuge. Alternatively, the initial fractionation, primarily on the basis of size, can be achieved by a standard differential centrifugation scheme (*see* Chapters 2, 3, and 5). Bands from the rate zonal gradient or pellets are then further subfractionated in Nycodenz™ gradients.

# 2. Materials

## 2.1. Homogenization

### 2.1.1. Monolayer Cells Using Isoosmotic Conditions

1. Tight-fitting 5-mL Dounce homogenizer: Use the smallest volume homogenizer that is practical. It is important for reproducible cell disruption that the cell suspension should fill as much of the usable volume as possible.
2. Phosphate-buffered saline (PBS): $0.12M$ NaCl, 2.6 m$M$ KCl, 8.1 m$M$ $Na_2HPO_4$, 1.5 m$M$ $KH_2PO_4$, pH 7.4.
3. ST: $0.25M$ Sucrose, 10 m$M$ triethanolamine-acetic acid, pH 7.6.
4. STE: $0.25M$ Sucrose, 10 m$M$ triethanolamine-acetic acid, pH 7.6, 1 m$M$ EDTA.

### 2.1.2. Monolayer and Suspension Culture Cells in Hypoosmotic Media

1. Tight-fitting Dounce homogenizer (*see* Section 2.1.1., item 1).
2. PBS (*see* Section 2.1.1., item 2).
3. 1 m$M$ Sodium bicarbonate.
4. 200 m$M$ KCl, 10 m$M$ $MgCl_2$, $2M$ sucrose, 50 m$M$ Tris-HCl, pH 7.6.

### 2.1.3. Monolayer and Suspension Culture Cells by Glycerol-Loading

1. PBS (*see* Section 2.1.1., item 2).
2. $1.2M$ Glycerol, 5 m$M$ Tris-HCl buffer, pH 7.4.
3. $0.25M$ Sucrose, 5 m$M$ Tris-HCl buffer, pH 7.4.
4. 50-mL Plastic tubes for low-speed centrifuge.
5. 50-mL Plastic syringe fitted with a 1-mm metal cannula.
6. Vortex mixer.

### 2.1.4. Nitrogen Cavitation

1. Nitrogen pressure vessel, models are manufactured by Baskerville Scientific Ltd., Manchester, UK and Artisan Industries, Waltham, MA.
2. PBS (*see* Section 2.1.1., item 2).
3. $0.25M$ Sucrose, 0.2 m$M$ $MgCl_2$, 5 m$M$ Tris-HCl, pH 7.4.
4. Oxygen-free nitrogen.
5. Plastic beaker, 50–100 mL.
6. Magnetic stirrer.

## 2.2. Subcellular Fractionation

1. Loose-fitting Dounce homogenizer (volume will vary with sample size).
2. Suspension medium (SM): $0.25M$ sucrose, 10 m$M$ HEPES-NaOH, pH 7.4.

3. Stock gradient solutions (in 10 m$M$ HEPES-NaOH, pH 7.4):
   a. 2.0$M$ Sucrose (55.5% [w/w]);
   b. 50% (w/v) Nycodenz™ (Nycomed Pharma AS); use the same buffer as diluent for gradient solutions.
4. Stock additives (in 10 m$M$ HEPES-NaOH, pH 7.4):
   a. 100 m$M$ MgCl$_2$;
   b. 100 m$M$ EDTA;
   c. 1$M$ KCl.
5. Polycarbonate or polyallomer centrifuge tubes for low-speed, high-speed, and ultracentrifuge rotors (both fixed-angle and swing-out). Volumes will vary with sample size, but for homogenates derived from $10^8$–$10^9$ cells, use 30- to 50-mL tubes in low-speed and high-speed rotors and 15- to 40-mL tubes for ultracentrifuge rotors.
6. Syringe (10–20 mL) fitted with wide bore (0.8 mm) metal cannula.
7. Gradient making device—optional (*see* Note 1).
8. An unloading device for swing-out rotor tubes. If the separated bands in a gradient are well resolved, then a Pasteur pipet whose tip is fashioned into an L-shape or a syringe and metal cannula (0.5 mm id) can be used for aspirating each band. If the bands are indistinct, use an upward displacement unloader (Nycomed Pharma AS).
9. Phase-contrast microscope.

# 3. Methods

## 3.1. Homogenization

### 3.1.1. Monolayer Cells Using Isoosmotic Conditions
(see *Note 2*)

Carry out all operations at 0–4°C. The method is applicable to five 9-cm dishes of confluent baby hamster kidney cells.

1. Remove the culture medium by aspiration, and wash each monolayer twice with 5 mL of PBS.
2. Wash each monolayer twice with 5 mL of ST.
3. Add 1 mL of STE (*see* Note 3) to each monolayer, and detach the cells from the substratum by scraping with a rubber policeman.
4. Use the policeman to transfer each of the crudely suspended monolayers to a small beaker.
5. Add another 1 mL of STE to each dish, and repeat the process to recover as many of the cells as possible. Avoid using a Pasteur pipet to transfer the material since the small sheets of cells tend to stick to the pipet surface and are difficult to recover.
6. Transfer half the suspended cells to the homogenizer and disrupt the cells using about 10 vigorous strokes of the pestle. Check for cell break-

age under the phase contrast microscope and continue the process until at least 95% rupture has been achieved, but do not use more than 20 up-and-down passes of the pestle (*see* Notes 4–7).
7. Repeat with the other half of the suspension.

### 3.1.2. Monolayer and Suspension Culture Cells in Hypoosmotic Media

1. Wash the cells at least three times in PBS. Suspension culture cells should be centrifuged at 200–500g for 5–10 min, depending on the cell type. After decantation or aspiration of the supernatant, resuspend the pellet in PBS (at least 10 vol of PBS to 1 vol of cell pellet). For mono-layer cells, *see* Note 8.
2. Resuspend the cells by the most gentle means possible (inversion, stir-ring with a glass rod, or careful aspiration into and from a Pasteur pipet). It is often easier to resuspend the cells in 1–2 mL of liquid before dilution to the appropriate volume.
3. Repeat the centrifugation. For cells that tend to aggregate when pelleted, 1 m$M$ EDTA may be added to the PBS.
4. After the final washing, suspend the cells in 1 m$M$ sodium bicarbonate ($10^8$ cells/cm$^3$), and after about 1 min on ice, transfer the suspension to the homogenizer (*see* Notes 9 and 10).
5. After 10 up-and-down strokes of the pestle, check the degree of homoge-nization by phase contrast microscopy and continue the process if required, up to about 25 strokes (*see* Notes 7 and 11).
6. Add 1/10 of the sucrose, KCl, MgCl$_2$, and Tris medium.

### 3.1.3. Monolayer and Suspension Culture Cells by Glycerol-Loading (see Note 12)

This method is applicable to hamster embryo fibroblasts (NIL-1), other cells may require adaptions (*see* Notes 7 and 13–16).

1. At confluence, scrape monolayers into PBS using a rubber policeman.
2. Wash the cells (approx $10^9$) in PBS three times; harvest the cell pellet each time by centrifugation at 500g for 5 min.
3. Resuspend the cells in 40 mL of the glycerol medium, and stir for 10 min at 4°C to allow them to take up the glycerol.
4. Centrifuge at 1000g for 5 min, and decant the supernatant.
5. Resuspend the cells in the residual supernatant using a vortex mixer, and add about 30 mL of the 0.25$M$ sucrose from the syringe (*see* Note 17).
6. Check that the cells have lysed by phase contrast microscopy.

### 3.1.4. Nitrogen Cavitation (see Note 18)

1. Set the nitrogen pressure vessel in ice to precool it.
2. Harvest the cells as normal (the precise method will depend on the cell type). Detach monolayer cells, by scraping with a rubber policeman, into PBS after washing the monolayer several times with the same medium. Sediment suspension culture cells at 200g for 10 min and wash twice with PBS (*see* Sections 3.1.1. and 3.1.2.).
3. Sediment the cells at 200g for 10 min, and resuspend in the sucrose medium. A cell concentration of about $2 \times 10^7$/mL is acceptable, but volumes <15 mL are difficult to handle efficiently (*see* Note 19).
4. Cool the suspension to 0°C, and transfer it to the plastic beaker within the nitrogen pressure vessel (*see* Note 19).
5. Assemble the pressure vessel according to the manufacturer's instructions, and place it within an ice bath on a magnetic stirrer (*see* Note 20).
6. While the suspension is stirred introduce oxygen-free nitrogen to a pressure of 5516 kPa (*see* Note 21).
7. Maintain the stirred cell suspension under these conditions for 20 min, and then with the delivery tube leading to a beaker open the delivery valve very slowly, keeping the tip of the delivery tube above the level of collected homogenate in the beaker.
8. As soon as all the suspension has been delivered and gas starts to emerge, close the delivery valve and vent the remaining gas.
9. Stir the homogenate gently to allow the foaming to subside before processing further.

## 3.2. Subcellular Fractionation of Homogenate

### 3.2.1. Differential Centrifugation

1. Centrifuge the homogenate at 1000g for 5 min; pour off and retain the supernatant and resuspend the pellet in SM (supplemented with the same additives as in the homogenization medium) to the original volume. Use only gentle agitation or repeated aspirations into a syringe, fitted with a wide bore cannula (*see* Note 22).
2. Repeat the centrifugation and washing process twice; bulk the supernatants together and centrifuge at 3000g for 10 min. Resuspend the pellet in SM (*see* Note 23).
3. Pour off and retain the supernatant; resuspend the pellet in SM to original homogenate volume using two or three gentle strokes of the pestle of the homogenizer.
4. Repeat the centrifugation; combine the supernatants, and resuspend the pellet in SM (*see* Note 23).

5. Continue this process at progressively higher relative centrifugal forces (rcf) (say 6000$g$ for 10 min, 10,000$g$ for 12 min, 20,000$g$ for 20 min, and 100,000$g$ for 30 min) to produce a series of resuspended pellets and a final cytosolic supernatant. The volumes used for resuspension can be reduced at each stage to make centrifugation more convenient.
6. Assay each resuspended pellet for relevant enzymes and for protein (*see* Notes 31–34 and Chapters 1 and 18).

### *3.2.2. Rate Zonal Separation in Sucrose Gradients* (see *Note 24*)

1. Adjust the concentration of sucrose in the homogenate to 0.55$M$ by adding the appropriate volume of 2$M$ sucrose, keeping the concentration of any additives the same.
2. Make a linear gradient of sucrose (0.73–1.9$M$), with any required additives, in tubes for an appropriate high-speed swing-out rotor, and layer the sample on top. The volume of sample should not exceed 15% of the volume of the gradient, e.g., a 20-mL sample could be loaded in 5-mL portions onto 4 × 40 mL gradients (*see* Note 25).
3. The centrifugations conditions must be adapted to suit a particular sample. Try 20,000$g$ for 1 h in pilot studies (*see* Notes 26 and 27).
4. Recover the banded material either using a Pasteur pipet or syringe and metal cannula (for well defined bands) or a gradient unloader (for poorly defined material).
5. Dilute the sample with at least an equal volume of 10 m$M$ HEPES buffer, and sediment the membranes by centrifugation at an appropriate speed (*see* Note 28).
6. Carry out the appropriate analyses (*see* Notes 31–34 and Chapters 1 and 18).

### *3.2.3. Separation in Nycodenz Gradients* (see *Note 29*)

1. To any pelleted material, either from the differential centrifugation of the homogenate or from the sucrose gradient, add 4 mL of 40% (w/v) Nycodenz and suspend it thoroughly using a Dounce homogenizer.
2. Adjust the concentration of the Nycodenz to 30% (w/v) by addition of buffer if necessary (refractive index of 30% Nycodenz = 1.3824).
3. Make up discontinuous gradients of 2.5 mL each of 10, 15, 20, 25, and sample 30% and 1 mL of 40% (w/v) Nycodenz, in suitable tubes for a swing-out rotor of an ultracentrifuge (*see* Note 30).
4. Centrifuge at 160,000$g$ for 3 h (*see* Note 30).
5. Harvest the banded material either using a gradient unloader, or if the bands are sufficiently distinct and separated using a Pasteur pipet or syringe.

6. Dilute each fraction with an equal volume of buffer and centrifuge at an appropriate rcf and resuspend in 0.25$M$ sucrose prior to assay (*see* Notes 8, 31–34 and Chapters 1 and 18).

## 4. Notes

1. Linear gradients can be produced using a standard two-chamber gradient maker or by discontinuous gradients that are allowed to diffuse. A 40 mL 0.73–1.9$M$ linear gradient, for example, can be produced by layering 8 mL each of 0.73, 1.0, 1.3, 1.6, and 1.9$M$, and leaving them overnight at 4°C.
2. Method adapted from Marsh et al. *(2)*.
3. If the nuclei are very sensitive to EDTA, then this might be omitted from the medium. If, however, the EDTA is integral to the ease of cell disruption, stabilizing ions such as 20 m$M$ KCl and/or 1 m$M$ MgCl$_2$ can be added once the cells have been homogenized.
4. Never use a homogenizer that is too large for the volume of cell suspension. It should never be less than half full for efficient and reproducible homogenization. Always choose the smallest homogenizer that is practicable, even though this may mean having to process the cells in two or three batches.
5. Always monitor the progress of homogenization by phase contrast microscopy. Check not only for cell rupture, but also that the nuclei are undamaged, i.e., they should be a dark gray color and unswollen. Nuclei from tissue culture cells seem to be particularly fragile and the release of DNA will lead to aggregation and loss of subcellular material.
6. If the cells do not homogenize satisfactorily with this protocol do not increase the number of strokes of the pestle further; use an alternative method. The aim must be to homogenize all the cells with the minimum number of strokes of the pestle. The more the number of strokes required, the less likely it is for the method to be reproducible, and the more variable will be the degree of fragmentation of the membranes and membranous organelles. Those organelles released from cells ruptured during the first pass of the pestle will be subjected to the greatest comminution, whereas those from cells ruptured in later passes will be increasingly intact.
7. If cells are kept in continuous culture over long periods of time, occasionally check that the method is working satisfactorily. Late-passage cells can change their characteristics, especially their susceptibility to disruption.
8. For monolayer cells, detach the monolayer from the substratum by the method best suited to your cells: incubation with trypsin or EDTA, or scraping with a policeman. Once this has been done, wash the cells in PBS.

9. The aim should be to keep the homogenate:cell volume as low as possible: The higher the concentration of cytosolic proteins in the homogenate, the greater the osmotic protection they afford to the released organelles, prior to the addition of the buffered sucrose/ion mixture.

10. Addition to the medium of divalent or monovalent cations prior to homogenization in order to provide immediate protection to the nuclei when they are released is feasible, but it may make the cells less easy to homogenize. Different cell types will respond individually to different homogenization protocols.

11. If you have a large number of cells and need to homogenize them in a number of batches, it is important that each batch be exposed to the bicarbonate for the same length of time before homogenization. The cells should be maintained in PBS, and centrifuged down and resuspended in bicarbonate only in amounts that can be handled in one homogenization step.

12. Method is adapted from Graham and Sandall *(3)*.

13. The osmotic gradient across the plasma membrane that occurs when the glycerol-loaded cells are placed in the isoosmotic sucrose is sufficient to cause lysis of NIL cells, Other cell types may require either the glycerol-loading conditions or the subsequent lysis to be modified in order to achieve satisfactory cell breakage.

14. Lysis may be aided by gentle pipeting or stirring in the sucrose solution. If the cell swell but fail to rupture, gentle Dounce homogenization may be required (try five up-and-down strokes).

15. In other cases, it may be necessary to increase the time of incubation in the glycerol medium.

16. Some suspension cells, e.g., Lettre cells, have been shown to require far more severe conditions: $4.2M$ glycerol and $0.1M$ sucrose with additional Dounce homogenization.

17. The released nuclei may need to be stabilized by inclusion of low concentrations ($0.5$–$1.0$ m$M$) of divalent cations in the sucrose medium.

18. Method adapted from Graham. *(4)*.

19. The large internal volume of the pressure vessel means that small volumes are difficult to handle. Volumes between 20 and 100 mL should be placed within a plastic beaker, and centralized within the vessel by a plastic collar. Larger volumes may be placed directly in the vessel; smaller volumes should not be used.

20. In some of the earlier models, the suspension was stirred by an impeller.

21. The pressure required may vary with the cell type.

22. The nuclei in this pellet tend to be fragile, especially if hypoosmotic conditions have been used for the homogenization; be very careful when washing it.

23. For assays the material should be resuspended in SM to 1–2 mg protein/mL; for further processing the pellet may be suspended in media other than SM.

24. Method adapted from Graham and Coffey *(5,6)*.

25. If the homogenate has too large a volume to be conveniently handled by a rate-zonal gradient, use a 20,000$g$ for 20 min fraction, from which the nuclei have been previously removed, resuspended in a suitable volume.

26. These gradients should be able to resolve nuclei, mitochondria, fragments of plasma membrane, and most membrane vesicles. It may be necessary to modulate the rcf, the time of centrifugation, or both, to improve the resolution of the particle of interest.

27. Although these gradients are called "rate-zonal," some of the larger particles may approach their banding density, and only the more slowly sedimenting particles are resolved on a size basis, especially if the rcf is increased.

28. When harvesting material banded within a gradient for enzyme analysis, the gradient medium (particularly sucrose) must be diluted to reduce its density and viscosity so that the particles within it can be pelleted. Do not use a greater rcf than is necessary to achieve pelleting; i.e., for nuclear fractions, up to 5000$g$ (10 min); mitochondria, lysosomes, membrane fragments, and so forth up to 20,000$g$ (20 min); and vesicles up to 200,000$g$ (1 h); otherwise the pellets may be difficult to resuspend. If the material in the gradient is adequately concentrated, it may be possible to carry out an enzyme assay on a gradient fraction directly, so long as the sucrose can be diluted sufficiently by the assay medium so that the final concentration is 0.2$M$ or less.

29. Method adapted from Graham and Coffey *(6)*.

30. This separation can be carried out in any suitable rotor; it can be adapted to a vertical rotor, in which case the centrifugation time can be reduced to 1 h.

31. The analyses used should include those for nuclei (DNA), plasma membrane (Na$^+$/K$^+$-ATPase), mitochondria (succinate-INT reductase), and endoplasmic reticulum (NADPH-cytochrome c reductase).

32. Glucose-6-phosphatase is widely used as a marker for the smooth endoplasmic reticulum from rat liver, but its presence in tissue culture cells is not entirely convincing.

33. Assays for Na$^+$/K$^+$-ATPase in tissue culture cell membranes may require a higher concentration of ouabain than that recommended in Chapter 1.

34. The 5'-nucleotidase may not be so clearly limited to the plasma membrane as it is in rat liver. For more information, *see* refs. *5* and *6*.

# References

1. Graham, J. M. (1982) Preparation of plasma membrane from tissue culture cells, in *Cancer Cell Organelles, Methodological Surveys*, vol. 11 (Reid, E., Cook, G. M., W., and Morre, D. J., eds.), Ellis Horwood, Chichester, UK, pp. 342–358.
2. Marsh, M., Schmid, S., Kern, H., Harms, E., Male P., Mellman, I., and Helenius, A. (1987) Rapid analytical and preparative isolation of functional endosomes by free flow electrophoresis. *J. Cell Biol.* **104,** 875–886.
3. Graham, J. M. and Sandall, J. K. (1979) Tissue culture cell fractionation I. Fractionation of membranes from tissue culture cells homogenized by glycerol-induced lysis. *Biochem. J.* **182,** 157–164.
4. Graham, J. M. (1972) Isolation and characterization of plasma membranes from normal and transformed tissue culture cells, *Biochem. J.* **130,** 1113–1124.
5. Graham, J. M. and Coffey, K. H. M. (1979) Tissue culture cell fractionation II. Zonal centrifugation of Lettre cells homogenized by glycerol-induced lysis: subfractionation in sucrose gradients. *Biochem. J.* **182,** 165–172.
6. Graham, J. M. and Coffey, K. H. M. (1979) Tissue culture cell fractionation III. The fractionation of cellular membranes from [125]I-lactoperoxidase labeled Lettre cells homogenized by bicarbonate-induced lysis: resolution of membranes by zonal centrifugation and in sucrose and metrizamide gradients. *Biochem. J.* **182,** 173–1890.

# CHAPTER 10

# The Isolation of Membranes from Bacteria

## Robert K. Poole

## 1. Introduction

### 1.1. Background

Perhaps the most striking feature of bacterial membranes is their multifunctional nature *(1,2)*. Bacterial cytoplasmic membranes, for example, catalyze the reactions of respiratory and photosynthetic electron transfer and associated energy transduction, and contain numerous carriers for solute transport in and out of the cell. The cytoplasmic membrane of *Escherichia coli* is believed to contain more than 200 protein types, of which about 60 may be involved in transport functions. In addition, there are the many highly hydrophobic protein complexes that constitute the respiratory chains, and the multicomponent $BF_1$ and $BF_0$ ATPase complexes. Bacterial membranes are distinctive in lacking sterols, but contain hopanes, polyterpenoids that have a role in maintaining membrane rigidity. Although variable, the major lipids are phosphatidylglycerol and phosphatidylethanolamine in gram-positive bacteria, and phosphatidylethanolamine with smaller amounts of phosphatidylglycerol in gram-negative bacteria.

Bacterial cell membranes are part of a multilayered envelope. The inner layer is the cytoplasmic membrane, about 8 nm thick, which is in direct contact with the cytoplasm and consists of a phospholipid bilayer in which proteins are embedded. Exterior to this are layers whose nature and number are very different in gram-positive and gram-negative bacteria. In the former, there is a thick (15–80 nm) layer of

From: *Methods in Molecular Biology, Vol. 19: Biomembrane Protocols: I. Isolation and Analysis*
Edited by: J. M. Graham and J. A. Higgins Copyright ©1993 Humana Press Inc., Totowa, NJ

murein (a peptidoglycan network of acetylated polysaccharide, crosslinked by short peptides) and teichoic acids, which forms the external barrier. In gram-negative bacteria, the peptidoglycan is thinner (2 nm) and covered by an outer membrane (about 8 nm thick). The inner leaflet of the latter is similar to that of the cytoplasmic membrane, but the outer layer consists of lipopolysaccharides, which serve to exclude hydrophobic compounds. Passive diffusion of hydrophilic compounds is, however, allowed by porins that form relatively nonspecific hydrophilic protein channels through which solutes with $M_r$ less than about 700 pass readily. The space (<5 nm deep) between outer and inner membranes is the periplasm, which contains solute-binding proteins, hydrolytic enzymes, and some electron carriers.

This chapter is concerned largely with the preparation from both gram-positive and gram-negative bacteria of inner or cytoplasmic membranes. Those onion-like layered invaginations of the cytoplasmic membrane that have been called mesosomes, but which may be artifacts of electron microscope preparation procedures, are not covered, nor are the "extra membranes" found in certain *E. coli* strains *(3)*. Brief reference is made to the preparation of chromatophore membranes from photosynthetic bacteria and of outer membranes from gram-negative bacteria. The reader is referred to more specialist accounts for preparation of *Halobacterium* purple membrane *(4)* and gas vesicles *(5)*. A recent compendium of useful techniques for characterization of bacterial cell surface structures and components is provided by Hancock and Poxton *(6)*.

## *1.2. General Strategy*

### *1.2.1. Disruption*

The isolation of bacterial membranes requires breakage or weakening of the strength- and shape-conferring layers of the envelope to allow release of membranes prior to subcellular fractionation and isolation of the required membrane fraction. Subcellular fractionation is invariably carried out by differential or zonal centrifugation, the principles of which are discussed elsewhere *(7)* but the choice of breakage technique is greater. It will be determined by the type of bacterium, and practical matters such as the availability of equipment and the required speed and simplicity of the procedures.

Techniques for disruption or disintegration of bacterial cells fall into two major categories, chemical and mechanical; in addition, cells can sometimes be disintegrated by freezing and thawing (they are summarized in Fig. 1). Chemical or enzymic processes exploit the peculiar chemistry of bacterial envelopes and involve two steps, namely the conversion of cells into osmotically sensitive forms, and their controlled lysis *(8,9)*. Osmotically sensitive protoplasts (from gram-positive bacteria) and spheroplasts (from gram-negative bacteria) can in turn be obtained in two ways. The penicillin method involves exposure to the antibiotic of cells growing in the presence of an osmotic stabilizer such as sucrose. The cell outgrows its peptidoglycan layer and lyses when transferred to hypotonic media. In a more common and convenient method, treatment of harvested bacteria with lysozyme causes hydrolysis of $\beta$-1,4-linkages in the relatively rigid peptidoglycan with resultant lysis and loss of cell contents. However, if the digestion is carried out in a hypertonic medium, i.e., of relatively high osmotic pressure, protoplasts or spheroplasts are formed and stabilized. These structures, lacking all (protoplasts) or most (spheroplasts) of the more rigid envelope layers, are osmotically sensitive and lyse, if placed in hypotonic medium, to give small membrane fragments that form vesicles of predominantly the same orientation ("right-side out") as in the intact cell.

An alternative general approach is to break the cells by mechanical means, generally involving liquid or solid shear forces. These may be generated, for example, by pressure extrusion of a cell suspension through a small orifice in the fluid (French pressure cell or Ribi cell fractionator) or solid state (Hughes press or X-press), by ultrasonication of cells in suspension, explosive decompression, or violent agitation of cells in suspension with beads or an abrasive such as alumina or sand. In general, membrane preparations made using the French press or by sonication (the two most common methods) consist mostly of small, everted ("inside out") topologically closed vesicles. When bacteria are freeze-pressed (e.g., in the Hughes press), the solid shear forces tend to cause disruption by transverse cleavage of the cell, causing little comminution of the envelope, which can be recovered as large, open-ended fragments. Detailed description of the principles and methods of operation of the diverse and ingenious

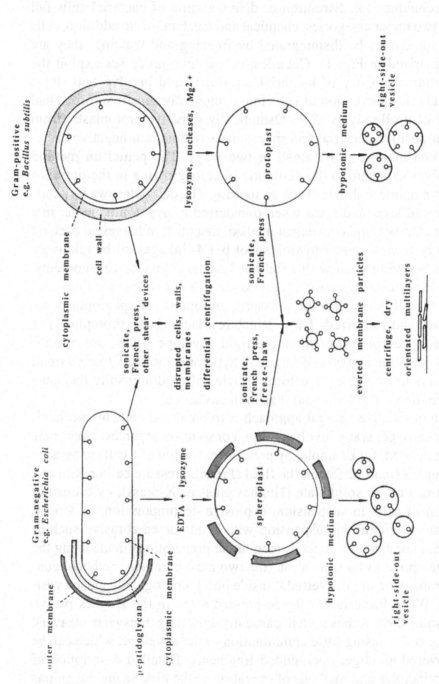

Fig. 1. Routes to the preparation of membrane vesicles and fragments from gram-negative and gram-positive bacteria. The stalk-like structures represent the ATPase molecules of the cytoplasmic membrane and, because they are directed inward in the intact cell, serve as indicators of the polarity of derived subcellular fractions.

devices available are outside the scope of this chapter but can be found in an excellent review *(10)*. In the sections that follow, details are provided of illustrative methods that are typical of those in common use.

### 1.2.2. Outer Membranes

Disruption of gram-negative bacteria followed by centrifugation yields a wall or envelope fraction in which some inner membrane remains attached to the peptidoglycan layer. Sometimes the peptidoglycan layer plus outer membrane ("cell wall") can be isolated free of inner membrane *(11)*, but isolation of pure outer membrane generally requires treatment with lysozyme to degrade the peptidoglycan enough to separate inner and outer membranes. The lysozyme treatment can be carried out before or after disruption. In the former, there is the risk that EDTA or the osmotic treatment necessary to expose the peptidoglycan may result in loss or reorganization of the outer membrane *(12)*. If the crude wall (plus membrane) fraction is treated with lysozyme, the two membranes can be separated by isopycnic density gradient centrifugation. The outer membrane usually, but not always *(13)* has the higher density. Density gradient flotation *(14)* may be used to improve the resolution of the two fractions.

### 1.2.3. Membrane Vesicles
### Prepared by Enzymic Cell Lysis

Membrane vesicles are widely used in studies of membrane transport. It is important to recognize that the polarity of topologically closed membrane vesicles (i.e., right-side out or inside out) and their suitability for transport studies is dramatically influenced by the choice of disruption method. The topography of vesicle preparations can lead to confusing results. Although those produced by lysozyme treatment are generally right-side out *(8)*, they can be heterogeneous *(15)*. Spheroplast preparation using lysozyme treatment is generally most effective with cells harvested in the exponential phase of growth, although Witholt et al. *(16)* have developed a procedure that renders variously grown *E. coli* completely susceptible to lysozyme.

### 1.2.4. Membrane Particles
### from Shear-Disrupted Cells

When the primary experimental objective is not the study of transport processes in vesicles with a specified polarity, but an investigation of the membranes' chemical or biochemical properties, it is generally preferable on the grounds of speed and yield to disrupt cells using, e.g., sonication or the French pressure cell, and recover membrane fragments and small vesicles by differential centrifugation. The procedures described here are equally suitable for most gram-positive or gram-negative bacteria.

### 1.2.5. Oriented Membrane Multilayers

Information about the orientation and arrangement of membrane-bound enzymes, and particularly of their active sites (e.g., the haem group of cytochromes), is especially desirable for enzymes such as respiratory chain components that catalyze transmembrane (vectorial) reactions. When mitochondrial membranes are centrifuged onto a flat surface and then partially dehydrated, they form multiple layers in which the planes of the membranes become well oriented, more or less parallel to each other and to the plane of the support *(17)*. When such multilayers are examined during rotation in the magnetic field of an ESR spectrometer or in the path of a plane-polarized light beam in a spectrophotometer, information can be deduced about the orientation of the plane of the haem relative to the plane of the membrane support. These studies were extended by Poole et al. *(18,19)* to examine the orientation of cytochrome haems, iron-sulfur clusters, and the molybdenum of nitrate reductase in membrane multilayers from *E. coli*. The method used in those studies is described in Sections 2.3. and 3.3.; membranes prepared this way are reasonably well oriented as revealed by electron microscopy *(18)*.

### 1.2.6. Chromatophores

Photosynthetic bacteria are often characterized by an extensive internal membrane system—the intracytoplasmic membrane. In many species this is thought to be contiguous with the cytoplasmic membrane. In species such as *Rhodospirillum rubrum*, *Rhodobacter* (formerly *Rhodopseudomonas*) *sphaeroides*, and *Rhodobacter capsulatus*, the intracytoplasmic membrane takes the form of invaginated and partly constricted tubules *(20)*. When the cells are subjected to mechanical disruption many of the tubules "pinch off" to form chromatophores, vesicular structures with diameters 10–40 nm. Because the intracytoplasmic membrane possesses the light-harvesting complexes, the photosynthetic reaction center, cytochrome $bc_1$ complex, and the ATP synthetase, the chromatophores are competent to perform photophosphorylation. The chromatophores are topologically inverted with respect to the intact cells, and so ADP and Pi are consumed and ATP is generated in the external medium.

Several procedures are available for mechanical disruption of the cells including grinding with alumina, exposure to ultrasonic sound, and passage through the French pressure cell. The chromatophores are separated from unbroken cells and cell debris by differential centrifugation *(21)*. If necessary, the membranes can be further fractionated by sucrose density gradient centrifugation. Most biochemical activities in chromatophores are stable for 1–2 wk if the preparations are stored at 4°C or for many months if the suspensions are supplemented with 50% glycerol and stored at –20°C.

## 2. Materials

### 2.1. Membrane Vesicles Prepared by Enzymic Cell Lysis

#### 2.1.1. Gram-Positive Bacteria

1. Appropriate volume of culture inoculated with organism of choice and grown preferably to exponential phase.
2. Preparative high-speed centrifuge capable of approx 25,000$g$.
3. Centrifuge bottles and tubes depending on culture volume.
4. Cell suspension buffer: 50 m$M$ potassium phosphate, pH 8.0.
5. Lysozyme, deoxyribonuclease-I and ribonuclease. The solids should be weighed out separately according to the volume of cell suspension anticipated (*see* Section 3.1.1.).
6. 0.5$M$ MgSO$_4$.
7. A shaking water bath or incubated shaker at 37°C.
8. 150 m$M$ potassium-EDTA in water or phosphate buffer, adjusted to pH 7.0.
9. Membrane suspension buffer: 0.1$M$ potassium phosphate, pH 6.6.
10. Hypodermic syringe, 10 or 20 mL, fitted with widest needle available (e.g., 18 g). Alternatively a small hand-held glass or Teflon homogenizer can be used for resuspending the membrane pellet.
11. Cryo-tubes or similar for storage in liquid nitrogen.

#### 2.1.2. Gram-Negative Bacteria

1. Appropriate volume of culture, inoculated with organism of choice and grown to exponential phase.
2. Preparative high-speed centrifuge capable of approx 46,000$g$.
3. Centrifuge bottles and tubes depending on culture volume.

4. Cell suspension buffer: 30 m$M$ Tris-HCl, pH 8.0 containing 50 µg/mL chloramphenicol and 20% (w/v) sucrose.
5. 100 m$M$ potassium-EDTA, adjusted to pH 7.0.
6. Solid lysozyme, preweighed to give a concentration of approx 0.5 mg/mL when added to the anticipated volume of cell suspension.
7. Spheroplast suspension buffer: 10 m$M$ potassium phosphate buffer, pH 6.6, containing 2 m$M$ MgSO$_4$, and DNase and RNase each at 10 µg/mL.
8. Hypodermic syringe, 10 or 20 mL, fitted with widest needle available (e.g., 18 g). Alternatively a small hand-held glass or Teflon homogenizer can be used for resuspending the membrane pellet.
9. Shaking water bath or incubated shaker at 37°C.
10. 1$M$ MgSO$_4$.
11. DNase, solid.
12. Storage buffer: 50 m$M$ potassium phosphate, pH 6.6.
13. Cryo-tubes or similar for storage in liquid nitrogen.

## 2.2. Membrane Particles from Shear-Disrupted Cells

1. Appropriate volume of culture inoculated with organism of choice and grown to exponential or stationary phase.
2. Preparative high-speed and ultra-centrifuge(s) capable of approx 12,000$g$ and 225,000$g$, respectively.
3. Appropriate centrifuge tubes and bottles, depending on culture volume.
4. Cell suspension buffer: 50 m$M$ Tris-HCl, pH 7.4 containing 2 m$M$ MgCl$_2$ and 1 m$M$ EGTA.
5. DNase, solid.
6. Probe ultrasonic disintegrator (150 or 200 W models are recommended).
7. Ice and salt bath.
8. Small hand-held homogenizer.
9. Cryo-tubes or similar for storage in liquid nitrogen.

## 2.3. Oriented Membrane Multilayers

1. A means of centrifuging membranes onto a flat surface normal to the direction of the centrifugal force is required. Thin-walled plastic tubes for a swing-out rotor can be fitted with matched tight-fitting hemispherical inserts machined from Perspex, on which is supported a circle cut from Mylar sheet, 0.1- to 0.2-mm thick. The sheet is washed first with detergent and water, and coated with collodion by immersing in a 1% (w/v) solution in ethanol/ether (75:25 [v/v]).
2. Membranes prepared by liquid shear, e.g., sonication. *See* Section 2.2.

# 3. Methods

## *3.1. Membrane Vesicles by Enzymic Cell Lysis*

### *3.1.1. Gram-Positive Bacteria*

A relatively simple procedure for preparing membrane vesicles is given by Bisschop and Konings *(22)*, modified from Konings et al. *(23)*. The latter paper details the ultrastructure and properties of such preparations.

1. Grow cells to late exponential phase in a defined or complex medium appropriate to the nutritional requirements of the particular bacterium used (*see* Note 1).
2. Harvest cells by centrifugation (e.g., 10,000*g* for 10 min), and suspend in 50 m*M* potassium phosphate buffer, pH 8.0, at 37°C to a concentration of 4 g wet wt per liter (*see* Note 2).
3. Add lysozyme, deoxyribonuclease-1, and ribonuclease as solids to final concentrations of 300, 10, and 10 µg/mL, respectively (*see* Note 3). After the enzymes have dissolved, add $MgSO_4$ as a small volume of stock solution (0.5*M*) to a final concentration of 10 m*M*.
4. Incubate for 30 min at 37°C.
5. Add potassium-EDTA to a final concentration of 15 m*M*, using the 150 m*M* stock solution.
6. Incubate for 1 min.
7. Add $MgSO_4$ solution to a final concentration of 10 m*M*.
8. Centrifuge at room temperature at 25,000*g* for 45 min.
9. Resuspend the membrane pellet in 0.1*M* potassium phosphate buffer, pH 6.6, at room temperature by means of a hypodermic syringe fitted with an 18-g needle to a concentration of 2–3 mg membrane protein per milliliter (*see* Note 4).
10. Distribute portions of suspension (e.g., 0.5 mL) in thin-walled plastic tubes, then freeze and store in liquid nitrogen. Before use thaw quickly in a water bath at 46°C (*see* Note 5).

### *3.1.2. Gram-Negative Bacteria*

A very detailed description of the preparation of spheroplasts from *E. coli* and other bacteria by lysozyme-EDTA or penicillin and their conversion into vesicles is given by Kaback *(8)*. The outline given here (*see* Note 6) is based on that account and the modifications described elsewhere *(24)*.

1. Harvest cells from an exponentially growing culture by centrifugation at 0°C, and suspend in 30 m$M$ Tris-HCl (pH 8.0) containing 20% sucrose and 50 µg chloramphenicol/mL.
2. Add, while stirring magnetically, potassium-EDTA (pH 7.0) and lysozyme to final concentrations of 10 m$M$ and 0.5 mg/mL respectively. Incubate for 30 min at room temperature.
3. Centrifuge the preparation at 13,000$g$ for 30 min to sediment the spheroplasts.
4. Disperse about 2 g (wet wt) in 10 mL of 10 m$M$ potassium phosphate buffer (pH 6.6) containing 2 m$M$ MgSO$_4$ and DNase and RNase at 10 µg/mL each. Disperse rapidly using a syringe and 18-g needle (*see* Section 3.1.1., item 9).
5. Incubate at 37°C for 30 min on a rotating shaker platform. If viscous after 15 min, add extra Mg$^{2+}$ (final 5 m$M$) and DNase (20 µg/mL) and continue incubation.
6. Centrifuge at 800$g$ for 1 h to remove whole cells and unlysed spheroplasts.
7. Carefully decant off supernatant, and centrifuge at 46,000$g$ for 1 h.
8. Suspend the membrane pellet in 50 m$M$ potassium phosphate buffer (pH 6.6) to a protein concentration of about 5–10 mg/mL. Store aliquots in liquid nitrogen.

## 3.2. Membrane Particles from Shear-Disrupted Cells

This procedure was described by Poole and Haddock *(25)* and has been used in this laboratory with minor modifications for many years.

1. Grow cells in a defined or complex medium; the phase of growth is of little importance for the success of the method (but of course has important physiological consequences).
2. Harvest cells by centrifugation (e.g., 12,000$g$ for 10 min), and wash in a buffer that contains 50 m$M$ Tris-HCl, 2 m$M$ MgCl$_2$, and 1 m$M$ EGTA, pH 7.4. This and all subsequent steps are at 4°C.
3. Resuspend the cell pellet in the same buffer; the weight of buffer used can be as little as equaling that of the wet cell pellet if a large quantity of cells is to be disrupted. Add a few flakes of DNase (*see* Note 7).
4. Sonicate the cell suspension in a hard glass tube immersed in an ice-salt slurry for cooling (*see* Note 8).
5. Centrifuge the suspension ("whole sonicate"; *see* Note 9) at 12,000$g$ for 15 min to yield a turbid supernatant fraction. Discard the pellet (*see* Note 10).
6. Centrifuge the sonicate supernatant at 225,000$g$ for 60 min (*see* Note 11) to give a reddish-brown pellet (often translucent; *see* Note 12) and

a yellowish supernatant solution. Retain the supernatant if desired, to check for distribution of marker enzymes.

7. Resuspend the pellet(s) in about the original volume of the same buffer using a small glass hand homogenizer to disperse all the material and make a uniformly turbid suspension. Centrifuge again at 225,000$g$ for 60 min.

8. Retain the supernatant if desired, and resuspend the pellet(s) in a minimal volume of the same buffer using a small homogenizer or syringe and needle (*see* Section 3.1.1., item 9). Dispense small aliquots into plastic tubes; freeze and store below –20°C, preferably in liquid nitrogen. *See* Note 13.

### 3.3. Oriented Membrane Multilayers

1. Prepare membrane particles from *E. coli* by ultrasonic disruption of cells and differential centrifugation (*see* Section 3.2.).

2. Dilute the thawed, concentrated membranes about 15-fold with distilled water, and homogenize in a hand homogenizer.

3. Prepare centrifuge tubes with inserts and Mylar supports as described in Section 2.3. Centrifuge in these tubes the membrane suspension at 70,000$g$ for 60 min in a swing-out rotor.

3. Cut open the centrifuge tube and extract the Mylar sheet. Drain off excess fluid, and incubate the film in an atmosphere of 90% relative humidity at 4°C for 48–72 h. For details of spectroscopic analyses of multilayers, refer to Section 1.2.5.

### 4. Notes

1. A strain of *Bacillus subtilis* was used by the authors.

2. We have found that this concentration can be increased 10-fold if necessary *(26)*.

3. DNase reduces the viscosity caused by the release of intracellular DNA. RNase and EDTA release and degrade RNA. $Mg^{2+}$ is required for DNase activity.

4. We favor the Markwell method *(27)* for protein determination, which tolerates high sucrose concentration (not used here) and readily solubilizes membrane suspensions (*see also* Chapter 18).

5. Ideally, use only membrane samples that have been frozen and thawed only once.

6. A similar, longer method is described by Kobayashi et al. *(28)*. Protoplasts are made by treatment with lysozyme and stabilized with glycylglycine or sucrose. They are transferred to a hypotonic medium containing deoxyribonuclease and $Mg^{2+}$. After homogenization the membranes are recovered by differential centrifugation and resuspended by passage through the French press before washing.

7. The function of the deoxyribonuclease is to degrade high-mol-wt DNA released from the cells, which increases the viscosity of the suspension, with concomitant reduction in cavitation and breakage efficiency.

8. For an MSE 150 W sonicator fitted with a probe having an end diameter of 9.5 mm, we have found five periods of 30 s each, interspersed with 15-s cooling intervals, satisfactory. The probe should be immersed by only a few millimeters into the sample, and the frequency adjusted to give a sizzling sound, reminiscent of frying eggs.

9. The sonication step can be replaced by two passages of the cell suspension through a French pressure cell (Aminco, Silver Spring, MD) at pressures around 100–150 MPa. The weight of buffer will have to be increased to at least twice that of the cell pellet.

10. The first pellet can be resuspended in buffer and repassaged through the disruption procedure if necessary.

11. Other combination of $g$ and time giving approximately the same integrated field-time are satisfactory.

12. If the first pellet obtained by centrifugation at high speed is suspended instead in a low ionic strength medium such as 1 m$M$ Tris-HCl (pH 7.4), the supernatant ("low ionic strength wash") resulting from the second such spin contains (in the case of *E. coli*; *25*) significant levels of dehydrogenase activity, the majority (96%) of F1-ATPase, and about 80% of the total prewash protein.

13. Variations of the general method involve the inclusion of protease inhibitors *(29)*, more extensive washing of the crude membranes *(30)* and the use of different centrifugation steps appropriate for other bacteria (e.g., *Azotobacter vinelandii*; *31*).

## Acknowledgments

I am particularly grateful to J. B. Jackson for his expert contribution to the section on chromatophores. Work in the author's laboratory in this area has been supported by the Royal Society, SERC, Smithkline and Beecham, and the Central Research Fund of the University of London.

## References

1. Cronan, J. E., Gennis, R. B., and Maloy, S. R. (1987) Cytoplasmic membrane, in *Escherichia coli and Salmonella typhimurium. Cellular and Molecular Biology* (Neidhardt, F. C., Ingraham. J. L., Low, K. B., Magasanick, B., Schaechter, M., and Umbarger, H. E., eds.), American Society for Microbiology, Washington, DC, pp. 31–55.

2. Neidhardt, F. C., Ingraham, J. L., and Schaechter, M. (1990) *Physiology of the Bacterial Cell.* Sinauer Associates, Sunderland, MA.

3. Greenawalt, J. W. (1974) The isolation of intracellular membranes of *Escherichia coli* 111a. *Methods Enzymol.* **31**, 633–642.
4. Oesterhelt, D. and Stoeckenius, W. (1974) Isolation of the cell membrane of *Halobacterium halobium* and its fractionation into red and purple membrane. *Methods Enzymol.* **31**, 667–678.
5. Walsby, A. E. (1974) The isolation of gas vesicles from blue-green algae. *Methods Enzymol.* **31**, 678–684.
6. Hancock, I. and Poxton, I. (1988) *Bacterial Cell Surface Techniques.* Wiley, Chichester, UK.
7. Lloyd, D. and Poole, R. K. (1979) Subcellular fractionation: isolation and characterization of organelles, in *Techniques in the Life Sciences, Biochemistry*, vol. B202 (Kornberg, H. L., ed.), Elsevier/North Holland, Amsterdam, pp. 1–27.
8. Kaback, H. R. (1971) Bacterial membranes. *Methods Enzymol.* **22**, 99–120.
9. Konings, W. N. (1977) Active transport of solutes in bacterial membrane vesicles. *Adv. Microbiol. Physiol.* **15**, 175–251.
10. Coakley, W. T., Bater, A. J., and Lloyd, D. (1977) Disruption of microorganisms. *Adv. Microbiol. Physiol.* **16**, 279–341.
11. Burnell, E., van Alphen, L., de Kruijff, B., and Lugtenberg, B. (1980) $^{31}$P nuclear magnetic resonance and freeze-fracture electron microscopy studies on *Escherichia coli* III. The outer membrane. *Biochim. Biophys. Acta* **597**, 518–532.
12. Lugtenberg, B. and van Alphen, L. (1983) Molecular architecture and functioning of the outer membrane of *Escherichia coli* and other Gram-negative bacteria. *Biochim. Biophys. Acta* **737**, 51–115.
13. Orndorff, P. E. and Dworkin, M. (1980) Separation and properties of the cytoplasmic and outer membranes of vegetative cells of *Myxococcus xanthus*. *J. Bacteriol.* **141**, 914–927.
14. Ishidate, K., Creeger, E. S., Zrike, J., Deb, S., Glauner, B., Macalister, T. J., and Rothfield, L. I. (1986) Isolation of differentiated membrane domains from *Escherichia coli* and *Salmonella typhimurium*, including a fraction containing attachment sites between the inner and outer membranes and the murein skeleton of the cell envelope. *J. Biol. Chem.* **261**, 428–443.
15. Futai, M. (1974) Orientation of membrane vesicles from *Escherichia coli* prepared by different procedures. *J. Membrane Biol.* **15**, 15–28.
16. Witholt, B., Boekhout, M., Brock, M., Kingma, J., van Heerikhuizen, H., and de Leij, L. (1976) An efficient and reproducible procedure for the formation of spheroplasts from variously grown *Escherichia coli*. *Anal. Biochem.* **74**, 160–170.
17. Erecinska, M., Blasie, J. K., and Wilson, D. F. (1977) Orientation of the hemes of cytochrome *c* oxidase and cytochrome *c* in mitochondria. *FEBS Lett.* **76**, 235–239.
18. Poole, R. K., Blum H., Scott, R. I., Collinge, A., and Ohnishi, T. (1980) The orientation of cytochromes in membrane multilayers prepared from aerobically grown *Escherichia coli* K12. *J. Gen. Microbiol.* **119**, 145–154.
19. Blum, H., Poole, R. K., and Ohnishi, T. (1980) The orientation of iron-sulphur clusters in membrane multilayers prepared from aerobically-grown *Escherichia coli* and a cytochrome-deficient mutant. *Biochem. J.* **190**, 385–393.

20. Hoh, S. C. and Marr, A. G. (1965) Location of chlorophyll in *Rhodospirillum rubrum. J. Bacteriol.* **89**, 1402–1412.
21. Clark, A. J., Cotton, N. J. P., and Jackson, J. B. (1983) The relation between membrane ionic current and ATP synthesis in chromatophores from *Rhodopseudomonas capsulata. Biochim. Biophys. Acta* **723**, 440–453.
22. Bisschop, A. and Konings, W. N. (1976) Reconstitution of reduced nicotinamide adenine dinucleotide oxidase activity with menadione in membrane vesicles from the menaquinone-deficient *Bacillus subtilis* AroD. Relation between electron transfer and active transport. *Eur. J. Biochem.* **67**, 357–365.
23. Konings, W. N., Bisschop, A., Veenhuis, M., and Vermeulen, C. A. (1973) New procedure for the isolation of membrane vesicles of *Bacillus subtilis* and an electron microscopy study of their ultrastructure. *J. Bacteriol.* **116**, 1456–1465.
24. Konings, W. N. and Kaback, H. R. (1973) Anaerobic transport in *Escherichia coli* membrane vesicles. *Proc. Nat. Acad. Sci. USA* **70**, 3376–3381.
25. Poole, R. K. and Haddock, B. A. (1974) Energy-linked reduction of nicotinamide adenine dinucleotide in membranes derived from normal and various respiratory-deficient mutant strains of *Escherichia coli. Biochem J.* **144**, 77–85.
26. James, W. S., Gibson, F., Taroni, P., and Poole, R. K. (1989) The cytochrome oxidases of *Bacillus subtilis*: mapping of a gene affecting cytochrome $aa_3$ and its replacement by cytochrome *o* in a mutant strain. *FEMS Microbiol. Lett.* **58**, 277–282.
27. Markwell, M. A. K., Haas, S. M., Bieber, L. L., and Tolbert, N. E. (1978) A modification of the Lowry procedure to simplify protein determination in membrane and lipoprotein samples. *Anal. Biochem.* **87**, 206–210.
28. Kobayashi, H., van Brunt, J., and Harold, F. M. (1978) ATP-linked calcium transport in cells and membrane vesicles of *Streptococcus faecalis. J. Biol. Chem.* **253**, 2085–2092.
29. Cox, G. B., Downie, J. A., Fayle, D. R. H., Gibson, F., and Radik, J. (1978) Inhibition by a protease inhibitor of the solubilization of the F1-portion of the $Mg^{2+}$-stimulated adenosine triphosphatase of *Escherichia coli. J. Bacteriol.* **133**, 287–292.
30. Senior, A. E., Fayle, D. R. H., Downie, J. A., Gibson, F., and Cox, G. B. (1979) Properties of membranes from mutant strains of *Escherichia coli* in which the β-subunit of the adenosine triphosphatase is abnormal. *Biochem. J.* **180**, 111–118.
31. Kelly, M. J. S., Poole, R. K., Yates, M. G., and Kennedy, C. (1990) Cloning and mutagenesis of genes encoding the cytochrome *bd* terminal oxidase complex in *Azotobacter vinelandii*: mutants deficient in the cytochrome *d* complex are unable to fix nitrogen in air. *J. Bacteriol.* **172**, 6010–6019.

CHAPTER 11

# Isolation and Purification of Functionally Intact Chloroplasts from Leaf Tissue and Leaf Tissue Protoplasts

### David G. Whitehouse and Anthony L. Moore

## 1. Introduction

### 1.1. Background

The isolation of photosynthetically active chloroplasts is the starting point for many plant metabolic studies as diverse as carbon assimilation, electron flow and phosphorylation, metabolite transport, and protein targeting. Whatever the subsequent use of the chloroplasts, it is preferable to attempt to isolate *intact* chloroplasts, as the organelle envelope will protect the stroma and thylakoids from the deleterious effects of degradative enzymes and phenolic compounds released from leaf tissues during the preparative procedures.

Intact chloroplasts can be isolated by either mechanical methods or derived from protoplasts. The choice of plant and method is largely dependent on whether the chloroplasts are for the study of photosynthesis itself or for investigation of the specific properties of a particular species. The voluminous literature on photosynthesis *per se* has been derived largely from studies using spinach chloroplasts (and to a lesser extent pea chloroplasts) isolated by the mechanical method.

From: *Methods in Molecular Biology, Vol. 19: Biomembrane Protocols: I. Isolation and Analysis*
Edited by: J. M. Graham and J. A. Higgins  Copyright ©1993 Humana Press Inc., Totowa, NJ

## 1.2. Experimental Strategy

Mechanical isolation is a fast and inexpensive method giving a high yield of chloroplasts but is restricted essentially to species with soft leaves having a low content of oxalate, phenolics, and starch. This usually means the use of pea (*Pisum sativum*) or spinach (*Spinacea oleracea*) leaves. By contrast, the protoplast method of isolation is a time-consuming and expensive method giving a low yield of chloroplasts, but it has the advantage of being applicable to the leaves of a wide range of plant species.

The principle of mechanical isolation involves homogenizing the leaf tissue to produce a "brei," which is filtered twice to remove large cell debris and unbroken tissue. The filtrate is subjected to a short centrifugation at relatively low speed followed by gentle resuspension of the pelleted chloroplasts. All operations are conducted at 0–4°C as speedily as possible.

Protoplasts can be prepared from a variety of plant species. The principle of preparation involves the digestion of the selected leaf material using a mixture of cell wall degrading enzymes (cellulase and pectinase) to free the protoplasts. Undigested leaf material is removed by filtration and the protoplasts collected by flotation in a gradient. Chloroplasts are easily obtained from the prepared protoplasts by disrupting the plasmalemma to release the cellular contents and organelles including the chloroplasts, which are then isolated by centrifugation.

This chapter describes both general and specific methods for the isolation and purification of intact chloroplasts. More detailed accounts and discussion of the methods described are to be found in Hall and Moore *(1)* and Walker *(2)*.

## 2. Materials

### 2.1. Mechanically Prepared Chloroplasts

1. Plants: For peas, Feltham First, Little Marvel, and Progress varieties have been used successfully. With spinach, two commonly used varieties are US hybrid 74 (also known as Yates hybrid 102) and Virtuosa. Sow peas in vermiculite to produce seedlings approx 5–6 cm tall after 10–14 d growth at 20–25°C with 9–12 h light of intensity >50 Wm$^{-2}$. Leaves of spinach should be taken from actively growing, nonflowering plants. Grow the plants in short days (8.5 h) under high light intensities (*see* Notes 1 and 2).

2. Isolation media: An effective medium for spinach is that of Cockburn et al. *(3)*: 330 m$M$ sorbitol, 10 m$M$ Na$_2$P$_4$O$_7$, 5 m$M$ MgCl$_2$, and 2 m$M$ Na-isoascorbate adjusted to pH 6.5 with HCl. Pyrophosphate media are not suitable for use with peas. Instead use a phosphate-containing medium: 330 m$M$ glucose, 50 m$M$ Na$_2$ HPO$_4$, 50 m$M$ KH$_2$PO$_4$, 5 m$M$ MgCl$_2$, 0.1% (w/v) BSA, 0.1% (w/v) NaCl, and 0.2% (w/v) Na-isoascorbate adjusted to pH 6.5 with KOH. The phosphate may be replaced by zwitterionic buffers, such as HEPES or MES (*see* Note 3).

3. Resuspension media: A general resuspension medium frequently used is that described in ref. *3*: 330 m$M$ sorbitol, 50 m$M$ HEPES-KOH pH 7.6, 2 m$M$ EDTA, 1 m$M$ MgCl$_2$, and 1 m$M$ MnCl$_2$. Where "low cation" isolation media have been used, it is desirable to also use them for resuspension.

4. Homogenizer: Motor-driven homogenizers, such as a kitchen blender, Polytron, Ultra-Turrax, or Waring blender, are preferable to a mortar and pestle.

5. Muslin and cotton wool.

6. Centrifuge tubes (50 mL) and either a bench or refrigerated low-speed centrifuge (*see* Note 4).

7. Two glass beakers (250-mL capacity).

8. Ice tray and ice.

9. Small soft paint brush for resuspending the chloroplast pellets.

10. Percoll™ (Pharmacia, Uppsala, Sweden or Sigma, St. Louis, MO).

## *2.2. Protoplast-Derived Chloroplasts*

1. Plants: Leaves from actively growing and healthy plants (*see* Note 5).

2. Carborundum powder, toothbrush, and single-edged razor blades.

3. Glass beaker (600-mL capacity).

4. Water bath at 25–30°C.

5. A 60–100 W lamp and glass trough with water (to act as a heat filter).

6. A nylon tea-strainer and nylon meshes of 20, 80, 200, and 500 µm.

7. Bench centrifuge and (10-20 mL) glass centrifuge tubes (*see* Note 6).

8. Digestive enzymes: Cellulase and pectinase (*see* Note 7).

9. Digestive mixture: There is no single digestion mixture that is suitable for all tissues and considerable trial and error may be needed before an optimal mixture is found. Cellulase is generally used in the range 1–3% (w/v), and pectinase in the range 0.1–0.5% (w/v). A good "starting" digestion mixture to try is: 2% (w/v) cellulase, 0.3% (w/v) pectinase, 500 m$M$ sorbitol, 5 m$M$ MES, and 1 m$M$ CaCl$_2$ adjusted to pH 6.0 with HCl (*see* Note 8).

10. Gradient media: Three ice-cold media are generally used.

Medium A: 500 m$M$ sorbitol, 5 m$M$ MES, and 1 m$M$ CaCl$_2$ adjusted to pH 6.0 with HCl.

Medium B: 500 m$M$ sucrose, 5 m$M$ MES (pH 6.0), and 1 m$M$ CaCl$_2$.

Medium C: 400 m$M$ sucrose, 100 m$M$ sorbitol, 5 m$M$ MES (pH 6.0), and 1 m$M$ CaCl$_2$.

11. Resuspension medium: As previously described (Section 2.1., item 3), it is suitable for many C3 and C4 systems.
12. To the end of a disposable syringe attach a 20-μm mesh using a plastic ring cut from a disposable pipet tip (*see* Note 9).

# 3. Methods
## 3.1. Mechanical Preparation of Chloroplasts

1. Harvest 50 g of leaves. The spinach must be deribbed before use, whereas the whole aerial shoot of the peas may be used rather than picking individual leaves. Rinse the leaves in ice-cold water, and shake to remove excess water.
2. Finely slice the leaf material, place in the chilled blender with 150 mL of an ice-cold slurry of isolation medium, and homogenize to a "brei" within 5–10 s with minimal frothing.
3. Squeeze the ice-cold brei through two layers of muslin into a 250-mL glass beaker on ice.
4. Pour the filtrate through a prewetted (with ice-cold medium) muslin-cotton wool sandwich (four layers of muslin either side of the cotton wool) into 250-mL glass beaker on ice (*see* Note 10).
5. Divide this filtrate equally between the centrifuge tubes, and centrifuge for 30–60 s, in the range 1500–6000$g$ (*see* Note 11).
6. Decant the supernatants, and with paper tissues remove the froth adhering to the sides of the tubes.
7. Remove the soft upper portion of the pellets (mostly broken thylakoids) by gentle swirling with ice-cold medium. Store the remaining pellets on ice until ready for resuspension (*see* Notes 12 and 13).
8. Add 0.5–1.0 mL of resuspension medium to each tube, and resuspend the pelleted chloroplasts by using a small paint brush or by gently shaking each tube.
9. Pool the chloroplasts suspensions, and store on ice.
10. To remove contaminants in the chloroplast suspension, e.g., mitochondria and catalase (from ruptured peroxisomes), resuspend the pellets, obtained at step 7, in 150 mL of ice-cold washing medium (either the isolation or resuspension media), and centrifuge and resuspend as before.
11. The percentage of intact chloroplasts may be enhanced (at the expense of yield) by using a simple step gradient with Percoll: Layer 4–6 mL of

chloroplast suspension onto a 6-mL cushion of 40% (v/v), Percoll containing osmoticum and buffer, and centrifuge at 2000–3000g for 1 min. Resuspend the pellet of intact chloroplasts as before (*see* Notes 13–16).

12. To obtain freshly prepared thylakoid membranes dilute the chloroplast suspension to 10% of its original strength; leave for 1 min and centrifuge as previously described (Section 3.1., item 5). Resuspend the pelleted, lysed thylakoids in full strength medium *(7)*.

## 3.2. Preparation of Protoplast-Derived Chloroplasts (see Note 19)

1. Expose the leaf tissue cells to the digestion mixture by (1) abrading the leaf surface with a toothbrush or carborundum, (2) stripping the epidermis off the leaf, or (3) slicing the leaf into transverse sections between 0.5- and 1-mm thick with a single-edged razor blade. Place the abraded or stripped surfaces in contact with the digestion mixture. It may be necessary to cut sections under 0.5$M$ sorbitol, drain, and then place in the digestion mixture. Alternatively use vacuum infiltration to aid exposure of the cell walls to the actions of the digestive enzymes.

2. Place 30 mL of the digestion mixture into a 600-mL capacity glass beaker, add the prepared leaf tissue (approx 3–5 g), and leave to digest at 25°C for 2–3 h with illumination provided by a 60- to 100-W bulb approx 20 cm above the digest, and a heat filter between the digest and lamp. After incubation the digest will contain chloroplasts, protoplasts, undigested tissue, and cell debris.

3. Decant the digestion mixture and store in a freezer for future reclamation of the enzymes. If the mixture contains appreciable numbers of protoplasts, recover these by centrifuging at 100–150g for 4–5 min (*see* Note 17).

4. Wash the digested tissue twice with 20 mL of medium A to liberate the protoplasts, and pour the washings through a 500-µm mesh (a nylon tea strainer can be used) followed by a 200-µm mesh.

5. Centrifuge the filtrate at 100–150g in glass centrifuge tubes for 4–5 min, and discard the supernatant.

6. Gently resuspend each pellet in its glass tube with approx 1 mL of medium B, and layer on to each 2 mL of medium C, this will act as a wash layer. Finally, complete the gradient in each tube with 1 mL of medium A. Ensure the gradient layers do not mix.

7. Centrifuge the tubes at 100–150g for 4–5 min, the intact protoplasts will float and may be collected at the interface between the two upper solutions using a Pasteur pipet (*see* Note 19). Dilute the recovered protoplasts approx 5- to 10-fold in medium A. For prolonged storage sucrose should replace the sorbitol.

8. A simpler gradient may be tried by adding 10% (v/v) Percoll to medium B containing the protoplasts, and overlay with 2 mL of medium A. Again collect the protoplasts, after centrifugation, at the interface and dilute 5- to 10-fold in medium A (*see* Notes 18 and 19).

9. If C4 plants are used follow the previous procedures (Section 3.2., steps 1–8), except filter the initial crude extracts through 500-μm and then 80-μm meshes. This results in the retention of bundle sheath protoplasts on the 80-μm sieve, and these may be resuspended in medium A. Harvest the C4 mesophyll protoplasts on the Percoll gradient since they are denser than C3 protoplasts (*see* Note 20).

10. Prepare intact chloroplasts from protoplasts by forcing them through a fine mesh, which will rupture the cell membrane and release the internal contents. Centrifuge the suspended protoplasts (200–400 μg chlorophyll) at 100–150$g$ for 2 min. Resuspend the pellets in ice-cold resuspension medium (Section 2.1., item 3) (*see* Note 16).

11. Using the prepared syringe (Section 2.1., item 12), draw up the resuspended protoplasts into the syringe and expel through the mesh; repeat the procedure once more (*see* Note 21). Now centrifuge the resultant suspension at 200–300$g$ for approx 1 min, and resuspend the chloroplast pellets in ice-cold chloroplast resuspension medium by gently shaking the tubes. The surface of these pellets is often "stringy" (especially at alkaline values) and may be removed with resuspension medium and discarded, leaving a highly intact chloroplast preparation (more that 95% intact) (*see* Note 15).

## 4. Notes

1. Healthy and actively growing plants are an essential prerequisite to obtaining metabolically active intact chloroplasts. Avoid high levels of starch (as the grains can rupture the chloroplast envelope during centrifugation) by harvesting plants at the start of the light period.

2. "New Zealand" and "Perpetual" spinach are not suitable for intact chloroplast preparation, as they are beet cultivars and contain high levels of oxalate and phenolics.

3. Several authors have found that media low in cations result in improved chloroplast intactness. Nakatani and Barber *(4)* originally developed this idea to enhance the separation of intact and broken chloroplasts: 330 m$M$ sorbitol, 20 m$M$ MES, and 0.2 m$M$ MgCl$_2$ adjusted to pH 6.5 with Tris. A variation of this "low cation" medium without Tris was reported by Cerovic and Plesnicar *(5)*: 340 m$M$ sorbitol, 2 m$M$ HEPES, 0.4 m$M$ KCl, and 0.04 m$M$ EDTA adjusted to pH 7.8 with KOH. Both of these media are suitable for use with peas. Other media may be prepared by substituting KCl (200 m$M$) for the sugar/sugar alcohol osmoticum *(6)*.

4. New and/or unscratched centrifuge tubes should be used as their walls will be smooth and will not present a roughened surface, which could rupture the chloroplast envelope during centrifugation. These tubes should be kept solely for chloroplast preparative work and cleaned and rinsed with distilled water only. A swing-out rotor is recommended, an unrefrigerated bench centrifuge can be used if the tubes and holders are prechilled, otherwise use a refrigerated centrifuge.

5. Although mature leaves of spinach and peas are suitable for protoplast preparation, generally young leaf tissues (1–2 wk old) give the best yields.

6. All glassware should be cleaned and rinsed with distilled water only and should be kept solely for protoplast preparative work. Use glass centrifuge tubes (for ease of viewing) when making the gradients.

7. The expensive digestive enzymes can be reclaimed by freezing discarded digestion mixtures; thawing and pooling these when appropriate, and precipitating the enzymes with ammonium sulfate. Freeze dry the precipitates for storage.

8. The "starting" digestion medium has proved successful with barley, pea, spinach, and wheat leaves.

9. Widen pipet tips (by cutting short the tip) to accommodate the diameter of the protoplasts (30–40 μm).

10. The muslin-cotton wool sandwich retains nuclei and broken chloroplasts, and fibrous ("hairy") muslin is best used for this.

11. Where a range of g forces and times are quoted, these will need to be determined for optimal results with each tissue used.

12. The presence of white ring(s) in the pellets indicates the presence of calcium oxalate and/or starch and shows the use of poor quality plants (which will only yield poor quality chloroplasts). A new source of plant material should be sought and used.

13. Storing the chloroplasts as pellets will prevent leaching of essential ions, such as potassium, into the medium.

14. Good quality intact chloroplasts may be expected to fix $CO_2$ at rates of approx 100 μmol/(mg chlorophyll)/h and above when they are illuminated in the appropriate media *(6)*.

15. Chloroplast intactness is assessed by comparing the rates of ferricyanide reduction before and after osmotically shocking the chloroplast suspension *(7)*. The ferricyanide will react only with thylakoids, and this forms the basis of the assay that is carried out in an oxygen electrode. Set up the following mixture adding the reagents in the order given:

    a. 1 mL Double-strength resuspending medium.

    b. 0.9 mL Water.

    c. 0.1 mL Chloroplast suspension (50–100 μg chlorophyll).

d. 10 μL 0.5*M* Potassium ferricyanide.

e. 10 μL 2*M* D,L-glyceraldehyde (this reagent inhibits $CO_2$ fixation and prevents possible interference).

Illuminate (using a projector or 150-W bulb, with a heat filter) this mixture in the electrode, and after approx 1 min add 10 μL 0.5*M* $NH_4Cl$ to uncouple electron flow and give rates of oxygen evolution not retarded by the possible build-up of a proton gradient. This gives the rate of ferricyanide reduction (*A*) by the intact chloroplast suspension. Repeat the assay with the order of addition of reagents as follows: Reagent (b) followed by (c), (a), (d), and (e), illuminate and add uncoupler as above. This assay will give the rate of reduction (*B*) by the fully broken chloroplasts. The percentage intactness of the original suspension is then calculated as:

$$\text{Intactness \%} = [(B - A)/B] \times 100$$

16. Chlorophyll content of the chloroplasts can be measured using the following method adapted from Bruinsma *(8)*: Add sufficient water to 10–100 μL to give a final volume of 1 mL followed by 4 mL of acetone. Mix and centrifuge at approx 3000*g* for 2–3 min. Read the absorbance of the supernatant at 652 nm (1 cm path-length), and multiply this value by 27.8 to give micrograms of chlorophyll per milliliter of the acetone extract. Take care to avoid evaporation or exposure to bright light of the acetone extract as these factors will introduce errors caused by concentration or photooxidation effects.

17. Exact g forces and centrifugation times need to be determined for optimal results with each tissue used, as these parameters can vary considerably for different plants.

18. Protoplasts from C4 plants are denser than C3, and the isolation gradient should be altered to 0.6*M* with respect to sucrose and sorbitol or Percoll used as described.

19. Isolated protoplasts are very fragile, and all pipeting and centrifugation operations must be carried out with care.

20. Protoplast intactness may be assessed by their ability to exclude Evans blue *(9)* and/or accumulate fluorescein *(10)*. The yield may be assessed by counting in a haemocytometer.

21. Prior to use, new syringes must be rinsed with alcohol to remove lubricant, which is inhibitory to photosynthesis.

## References

1. Leegood, R. C. and Walker, D. A. (1983) Chloroplasts, in *Isolation of Membranes and Organelles from Plant cells* (Hall, J. L. and Moore, A. L., eds.), Academic, New York, pp. 185–210.

2. Walker, D. A. (1980) Preparation of higher plant chloroplasts, in *Methods in Enzymology*, vol. 69 (San Pietro, A., ed.), Academic, New York, pp. 94–104.

3. Cockburn, W., Walker, D. A., and Baldry, C. W. (1968) Photosynthesis by isolated chloroplasts. *Biochem. J.* **107,** 89–95.

4. Nakatani, H. Y. and Barber, J. (1977) An improved method for isolating chloroplasts retaining their outer membranes. *Biochim. Biophys. Acta* **461,** 510–512.

5. Cerovic, Z. G. and Plesnicar, M. (1984) An improved procedure for the isolation of intact chloroplasts of high photosynthetic capacity. *Biochem. J.* **223,** 543–545.

6. Robinson, S. P. (1986) Improved rates of $CO_2$-fixation by intact chloroplasts isolated in media with KCl as the osmoticum. *Photosynthesis Research* **10,** 93–100.

7. Lilley, R. McC., Fitzgerald, M. P., Reinits, K. G., and Walker, D. A. (1975) Criteria of intactness and the photosynthetic activity of spinach chloroplast preparations. *New Phytol.* **75,** 1–10.

8. Bruinsma, J. (1961) A comment on the spectrophotometric determination of chlorophyll. *Biochim. Biophys. Acta* **52,** 576–578.

9. Nagata, H. and Takebe, I. (1970) Cell wall regeneration by isolated tomato fruit and protoplasts. *Protoplasma* **64,** 460–480.

10. Power, J. B. and Chapman, J. V. (1985) Isolation, culture and genetic manipulation of plant protoplasts, in *Plant Cell Culture: A Practical Approach* (Dixon, R. A., ed.), IRL, Oxford, pp. 37,38.

# CHAPTER 12

# Isolation and Purification of Functionally Intact Mitochondria from Plant Cells

*Anthony L. Moore, Anne-Catherine Fricaud,*
*Andrew J. Walters, and David G. Whitehouse*

## 1. Introduction

The isolation of mitochondria from plant cells that display similar biochemical and morphological characteristics to those observed in vivo requires considerable expertise. Although it is relatively easy to prepare a crude mitochondrial fraction by differential centrifugation, marked changes in functional capabilities and morphological appearance are often observed in such fractions, suggesting that some degree of structural damage has occurred during the isolation procedure. Most of the problems involved in the isolation of intact and fully functional mitochondria occur during the homogenization phase because of the high shearing forces required to rupture the plant cell wall. Such forces tend to have a deleterious effect on other subcellular organelles, such as the vacuole, resulting in the release of degradative enzymes and secondary products, such as flavonoids and phenolic compounds, which may severely impair mitochondrial integrity. Thus the effects of homogenization on mitochondrial structure and function range from undesirable to totally destructive. Obviously techniques that reduce all or any of these problems are desirable.

The purpose of this chapter is to describe not only general considerations in the isolation of plant mitochondria, but also to suggest specific techniques for the isolation and purification of functionally

From: *Methods in Molecular Biology, Vol. 19: Biomembrane Protocols: I. Isolation and Analysis*
Edited by: J. M. Graham and J. A. Higgins Copyright ©1993 Humana Press Inc., Totowa, NJ

active and intact mitochondria from etiolated and green leaf tissue. Unfortunately, there is no accepted standard method for the isolation of mitochondria from plant tissues, and each tissue presents its own difficulties. Difficulties can be minimized, however, by careful choice of the plant material. For a detailed analysis of specific techniques employed for the isolation and purification of mitochondria from specific tissues, the reader is referred to Douce et al. *(1)*, Moore and Proudlove *(2,3)*, Edwards and Gardestrom *(4)*, and Neuburger *(5)*.

## 2. Materials

1. Isolation medium: $0.3M$ mannitol, 7 mM L-cysteine, 2 mM EDTA, 0.1% (w/v) defatted bovine serum albumin (BSA), 0.6% (w/v) polyvinyl-pyrrolidone (PVP-40, $M_r$ 40,000), and 40 mM 3-($N$-morpholino)-propane-sulfonic acid (MOPS) adjusted to pH 7.5 with $10M$ KOH. In the case of green leaf mitochondria, mannitol is replaced by $0.3M$ sucrose, L-cysteine by 3 mM β-mercaptoethanol, the BSA concentration is increased to 0.4% (w/v), PVP-40 to 1% (w/v) and 1 mM $MgCl_2$ is included.
2. Washing medium: $0.3M$ mannitol, 2 mM EDTA, 0.1% (w/v) BSA, and 40 mM MOPS adjusted to pH 7.4. With green leaf tissue replace mannitol with sucrose.
3. Purification medium (etiolated tissues): $0.3M$ sucrose, 0.1% (w/v) BSA, 5 mM MOPS, pH 7.4, and 22% (v/v) Percoll™ (Pharmacia, Uppsala, Sweden or Sigma, St. Louis, MO).
4. Purification medium (green leaf tissues): Solution A contains $0.3M$ sucrose, 0.1% (w/v) BSA, 10 mM phosphate buffer, pH 7.4, 10% (w/v) PVP-40, and 28% (v/v) Percoll. Solution B contains $0.3M$ sucrose, 0.1% (w/v) BSA, 10 mM phosphate buffer, pH 7.4, and 28% (v/v) Percoll.
5. Six 500-mL centrifuge bottles for a high-speed centrifuge.
6. Six 50-mL centrifuge tubes for a high-speed centrifuge.
7. A paint brush for resuspending the mitochondrial pellets.
8. A gradient maker (for green leaf tissue).
9. Domestic liquidizer, e.g., Moulinex Juice Extractor (Type 140) or hand-held Moulinex or Braun mixer.
10. Polytron or Ultra-Turrax homogenizer (for green leaf tissue only).
11. Cheesecloth (or muslin).
12. $10M$ KOH.

## 3. Methods
### 3.1. Etiolated Tissues (see Notes 1–4)

1. Peel and wash fresh material (1.5 kg) in cold water, and homogenize using a juice extractor (Moulinex type 140) or other mild method of

homogenization (hand-held Moulinex or Braun mixer). In the case of the juice extractor, collect the juice directly into 1.3 L of isolation medium maintained at 4°C. During this operation, maintain the pH at pH 7.6 by dropwise addition of 10$M$ KOH. If a hand-held homogenizer is used, dice the plant material directly into the isolation medium and homogenize by macerating for short periods (10–15 s) at 4°C. An outline of the procedure is given in Fig. 1 (*see* Notes 2 and 3).

2. Rapidly squeeze the brei through four layers of cheesecloth, and centrifuge the filtered suspension in 500-mL bottles at 400$g$ for 5 min at 4°C.
3. Discard the pellet, and centrifuge the supernatant at 3000$g$ for 10 min. Discard the pellet that contains nuclei, unbroken cells, and cell fragments.
4. Centrifuge the supernatant at 17,700$g$ for 10 min.
5. Following centrifugation, aspirate the supernatant and gently resuspend the pellet, which is highly enriched in mitochondria, in 3–4 mL of washing medium using a paint brush. Pool the resuspended pellets, dilute to approx 80–90 mL, and transfer the suspension to two 50-mL centrifuge tubes.
6. Harvest the washed mitochondria by centrifugation at 12,000$g$ for 10 min. Discard the supernatant, and gently resuspend the pellet in a small volume of wash medium and store at approx 50 mg protein/mL.
7. To prepare purified mitochondria, layer 2–3 mL of the washed mitochondria on 36 mL of the purification medium in two 50-mL tubes, and centrifuge at 22,300$g$ for 30 min (*see* Note 6).
8. Intact mitochondria are recovered as a broad band in the bottom third of the gradient below a yellowish band containing numerous small vesicles and above the peroxisomal pellet.
9. Collect the mitochondrial fraction, after careful aspiration of the upper part of the gradient, using a long-tipped Pasteur pipet. Dilute 10–15 times with washing medium to reduce the Percoll concentration.
10. Harvest purified mitochondria by centrifugation at 12,000$g$ for 15 min, and resuspend the pellet in minimum volume (0.5 mL) of wash medium.

### 3.2. Green Leaf Tissue (see Notes 1–4)

1. Grow pea plants from seed in vermiculite or Palmers pot bedding compost for 15 d under a 12-h photoperiod of warm white light from fluorescent tubes. Use approx 100 g of fully expanded leaves per 600 mL of isolation medium at 4°C.
2. Homogenize the leaves using a hand-held homogenizer for approx 30–45 s taking care to minimize frothing (*see* Notes 2 and 3). This procedure results in only partial maceration of the tissue.
3. Complete the disruption procedure by transferring approx 150-mL portions of the homogenate to a vessel suitable for use with a Polytron or Ultra-Turrax homogenizer and homogenizing for 3–5 s at full speed.

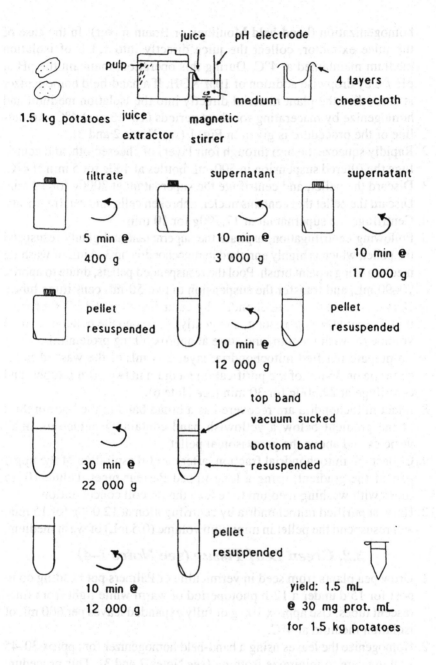

Fig. 1. A flow diagram for the isolation and purification of mitochondria from potato tubers.

4. Strain the homogenate through four layers of muslin (or cheesecloth) prewetted with isolation medium.
5. Transfer the filtrate into the 500-mL bottles, and centrifuge at 2960g for 5 min.
6. Discard the pellet, and centrifuge the supernatant at 17,700g for 15 min.
7. Resuspend the pellets gently in 3–4 mL of washing medium using a paint brush, and dilute each to 30 mL with washing medium.
8. Transfer the resuspended pellets into 50-mL tubes, and centrifuge at 1940g for 10 min.
9. Centrifuge the supernatant for 12,100g for 10 min to pellet the washed mitochondria, and resuspend in 2–3 mL of washing medium.
10. The procedure for the isolation of chlorophyll-free mitochondria from pea leaves employs a self-generating gradient of Percoll in combination with a linear gradient of PVP-40. Place 17 mL of solution A to the chamber of the gradient mixer proximal to the exit tube and 17 mL of solution B to the distal chamber. Agitate solution A with a stirrer, and immediately open the passage way between the two sides of the gradient maker and use a peristaltic pump to achieve a flow rate of approx 5 mL/min. Layer the incoming solution into a 50-mL centrifuge tube. Repeat the operation to obtain a second Percoll/PVP-40 gradient (*see* Note 6).
11. Carefully layer 2–3 mL of the resuspended washed mitochondrial pellet onto each of the gradients, and centrifuge at 39,200g for 40 min. Mitochondria are found as a tight light brown band near the bottom of the tube while thylakoids remain at the top of the tube.
12. Carefully remove the mitochondrial fraction, after aspiration of the upper part of the gradient, using a long-tipped Pasteur pipet, and dilute the suspension at least 10-fold with washing medium.
13. Harvest purified mitochondria by centrifugation at 12,100g for 10 min. Aspirate the supernatant taking care not to disturb the soft mitochondrial pellet. Add approx 30 mL of washing medium, then centrifuge the resuspension at 12,100g for a further 10 min.
14. Remove the supernatant by aspiration, and resuspend the purified mitochondrial pellet in minimum volume (0.5 mL) using a paint brush.

## 4. Notes

1. The procedures outlined are most suitable for tuberous tissue (1.5 kg/1.5 L) such as potato tubers, sweet potatoes, cauliflower buds, Jerusalem artichokes, beetroots, and turnips. If seedlings (3–400 g/1.5 L) such as mung bean hypocotyls or maize are used, then the tissue is cut with scissors directly into the isolation medium, which is made up without

PVP-40. If functionally specialized mitochondria are required (i.e., those containing a high degree of cyanide-insensitive alternative oxidase), then tissues such as the spadices (100 g/200 mL) of Lords and Ladies (*Arum maculatum*) or Voodoo Lily (*Sauromatum guttatum*) should be used. In these cases, the tissue is sliced directly into the isolation medium (in which the BSA is increased to 0.2% (w/v) and PVP-40 is omitted). Spinach (600 g/1.5 L) can be substituted for pea leaves. The leaves should be deribbed prior to use.

2. Mitochondria require an osmoticum in order to maintain structural integrity; sucrose, mannitol, or sorbitol are most commonly used. Acidity, phenolic compounds, tannin formation, and oxidation products rapidly inactivate mitochondria, hence it is important to ensure the buffers are alkaline and to add either PVP and cysteine (or β-mercaptoethanol) to the isolation medium. Note that all media are made up fresh just prior to the isolation procedure. BSA may be added as the last ingredient and left to dissolve without stirring.

3. The most crucial aspect of the whole procedure for the isolation of functionally intact mitochondria is the grinding of the tissue. In order to maintain structural integrity, the grinding procedure must be kept to an absolute minimum. Although longer blending improves the final yield, it drastically reduces the percentage of intact mitochondria. The key point is to sacrifice quantity for quality. Numerous techniques can be used to break the cell wall, but the ones outlined in the Methods section tend to be the mildest. A Waring blender can be used in place of a juice extractor or hand-held homogenizer, but disruption must be done at low speed for a maximum of 3–5 s. The blades should be routinely sharpened after several homogenizations to ensure that cutting rather than shearing of the tissue occurs. Care should also be taken to prevent excessive frothing, since this results in protein denaturation.

4. The fractionation procedure should be as rapid as possible in order to remove mitochondria from the degradative enzymes present in the extract.

5. With tuberous tissue, isolation procedures may result in the inactivation of succinate dehydrogenase. This can be overcome by the inclusion of 1 m$M$ succinate in the washing medium. Similarly, 2–5 m$M$ glycine is often included in the washing medium for green leaf tissue to prevent inactivation of glycine decarboxylase.

6. Although there are a number of other techniques available to purify mitochondria, such as on sucrose gradients (*6*) or by partition in aqueous polymer two-phase systems (*7*), the use of Percoll (a colloidal silica coated with polyvinyl-pyrrolidone) has proved to be the most successful. Centrifugation of Percoll-containing solutions results in the forma-

tion of a self-generated continuous gradient. Although this normally occurs at centrifugation speeds in excess of 40,000*g* (at 28% [v/v]) the protocol outlined is equally as effective at 22,000*g* (at 22% [v/v]). Thus it should be noted that a number of parameters influence the shape of the gradient, such as Percoll concentration, rotor geometry, diameter of tubes, centrifugal force, and time. Similarly choice of the osmoticum (mannitol, sucrose, or raffinose) can also modify the density and viscosity of the medium and a combination of these can prove effective at separating mitochondria from other contaminating membranes or organelles *(1)*. For details of the characteristics and behavior of Percoll, the reader is referred to the Pharmacia booklet entitled *Percoll Methodology and Applications*.

## References

1. Douce, R., Bourguignon, J., Brouquisse, R., and Neuburger, M. (1987) Isolation of plant mitochondria: General principles and criteria of integrity. *Methods Enzymol.* **18,** 403–415.
2. Moore, A. L. and Proudlove, M. O. (1983) Mitochondria and sub-mitochondrial particles, in *Isolation of Membranes and Organelles from Plant Cells* (Hall, J. L. and Moore, A. L., eds.), Academic, New York, pp. 153–184.
3. Moore, A. L. and Proudlove, M. O. (1987) Purification of plant mitochondria on silica sol gradients. *Methods Enzymol.* **18,** 415–420.
4. Edwards, G. E. and Gardestrom, P. (1987) Isolation of mitochondria from C$_3$, C$_4$ and crassulacean metabolism plants. *Methods Enzymol.* **18,** 421-434.
5. Neuburger, M. (1985) Preparation of plant mitochondria, criteria for mitochondrial integrity and purity, survival *in vitro*, in *Encyclopedia of Plant Physiology, Vol. 18: Higher Plant Cell Respiration* (Douce, R. and Day, D. A., eds.), Springer-Verlag, Berlin, pp. 7–24.
6. Douce, R., Christensen, E. L., and Bonner, W. D. (1972) Preparation of intact plant mitochondria. *Biochim. Biophys. Acta* **275,** 148–160.
7. Gardestrom, P. and Ericson, I. (1987) Separation of spinach leaf mitochondria according to surface properties: Partition in aqueous polymer two-phase systems. *Methods Enzymol.* **18,** 434-442.

## CHAPTER 13

# Isolation of Cholinergic-Specific Synaptosomes by Immunoadsorption

## J. Paul Luzio and Peter J. Richardson

### 1. Introduction

The most widely used subcellular fractionation techniques (e.g., centrifugation) are dependent on differences in physical parameters between organelles (e.g., size and density). In contrast, immunological techniques rely on biological differences, i.e., the expression of different antigens. The most popular immunological method for subcellular fractionation is immunoadsorption onto a solid phase, which offers the prospect of fast purification in high yield under conditions that do not expose organelles to unusual media or to osmotic shock. Immunoisolation techniques also have the advantage of being applicable to situations where only a small amount of cell material is available, e.g., tissue culture cells.

There has been a growing interest in the approach since the first reports of successful immunoaffinity purification of plasma membrane fractions (1–3). Several techniques have been employed and their relative merits reviewed (4–11). They can be divided into the use of either direct immunoadsorbents, where antiorganelle antibody is itself covalently bound to the solid phase, or indirect immunoadsorbents, where an antibody-binding reagent is covalently bound to the solid phase providing an immunoadsorbent capable of isolating the antibody-organelle complex (Fig. 1). In practice, direct immunoadsorbents have rarely proved successful and it has been argued, by analogy with affinity chromatography, that an indirect technique is required

From: *Methods in Molecular Biology, Vol. 19: Biomembrane Protocols: I. Isolation and Analysis*
Edited by: J. M. Graham and J. A. Higgins Copyright ©1993 Humana Press Inc., Totowa, NJ

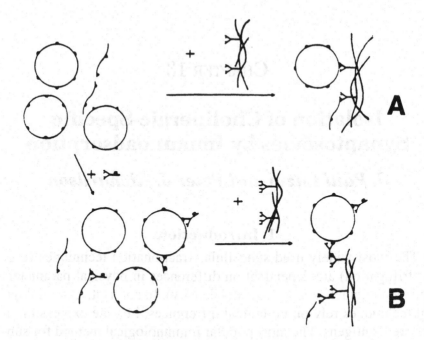

Fig. 1. Schematic representation of direct (**A**) and indirect (**B**) immunoaffinity techniques for the preparation of subcellular organelles and membranes. In the direct method (**A**) the homogenate or partially purified subcellular fraction is reacted with the antiorganelle antibody immobilized to give a solid phase immunoadsorbent (shown as a flexible, cellulose matrix). In the indirect method (**B**) the free antiorganelle antibody is reacted with the homogenate (or fraction), the unbound antibody washed away, and the organelles then isolated on an immunoglobulin-binding immunoadsorbent.

either to provide a more flexible spacer arm or for steric reasons owing to the difficulties inherent in the binding of two large particles.

The antibody-binding reagent coupled to the solid support may be either an antiimmunoglobulin second antibody (polyclonal or monoclonal) or proteins A or G, which bind to the $F_c$ portion of many immunoglobulin molecules. The indirect technique makes it possible to standardize the solid phase for use in the preparation of many organelles using different antibodies.

The choice of first and second antibodies may be dictated by availability, although it is often possible to incorporate some element of design into the choice. In principle, the first antibody should be specific to the organelle or membrane fragment of interest, be of high avidity, and directed to an antigen(s) with a density >50 mol/$\mu$m$^2$ *(12,13)*. The antiimmunoglobulin antibody coupled to the solid phase for indirect immunoaffinity purification should be of high avidity, react with many immunoglobulin subclasses, and be available in large amount. Monoclonal antibodies are particularly useful since essentially unlimited quantities of antibody can be produced from a single hybridoma cell line (*see Biomembrane Protocols: II. Architecture and Function,* Chapter 5).

The choice of solid support is itself determined by a number of criteria; the most important of these are high binding capacity, nonporous nature, adequate spacer arms, low nonspecific binding, ease of separation from unbound particles, and availability. In practice, the two most successful high capacity solid supports are cellulose fibers and nonporous, magnetic, monosized, polymer beads. The latter are marketed as Dynabeads™ by Dynal A/S (Oslo, Norway; Great Neck, NY; and Wirral, Merseyside, UK), with or without antiimmunoglobulin reagents covalently coupled.

The production of cellulose immunoadsorbents *(14)* is simple and cheap. They are easy to use in the separation of bound and free particles since this can be achieved by low-speed centrifugation. Many plasma membrane fractions have been prepared using cellulose immunoadsorbents and they have also been successfully applied to the preparation of membrane vesicles from intracellular membrane traffic pathways *(15,16)*. The chemistry of protein binding to cellulose is shown in Fig. 2 and we provide herein a protocol for the preparation of a cellulose immunoadsorbent *(14)* and for the immunoaffinity purification of cholinergic specific synaptosomes (nerve terminals) from rat cerebral cortex *(17)*. The immunoadsorbent is normally prepared well in advance of membrane isolation allowing the preparation of membrane fractions within 2 h of adding antimembrane antibody.

## 2. Materials

1. m-Nitrobenzyloxymethyl pyridinium chloride used in the preparation of aminocellulose from Sigma Chemical Co. (St. Louis, MO).
2. Cellulose powder (CC41) and starch iodide paper from Whatman Ltd., Maidstone, Kent, UK.
3. Sodium dithionite solution: 200 g/L prepared immediately before use.

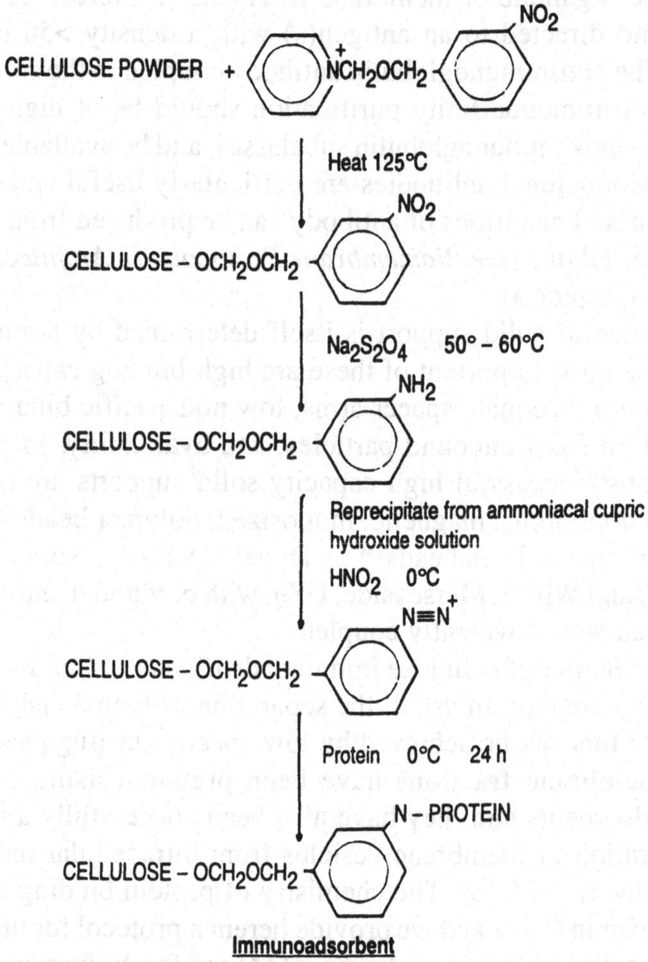

CELLULOSE POWDER + [ring]$\overset{+}{N}CH_2OCH_2$[ring]$NO_2$

Heat 125°C

CELLULOSE – $OCH_2OCH_2$[ring]$NO_2$

$Na_2S_2O_4$ 50° – 60°C

CELLULOSE – $OCH_2OCH_2$[ring]$NH_2$

Reprecipitate from ammoniacal cupric hydroxide solution

$HNO_2$ 0°C

CELLULOSE – $OCH_2OCH_2$[ring]$N\equiv\overset{+}{N}$

Protein 0°C 24 h

CELLULOSE – $OCH_2OCH_2$[ring]N – PROTEIN

**Immunoadsorbent**

Fig. 2. General scheme for preparation of cellulose-based immunoadsorbent.

4. Ammoniacal cupric hydroxide used in making diazocellulose, prepared immediately before use as follows: Dissolve 1.5 g of cupric chloride in a minimum amount of water (5 mL) to form a green solution, and add fresh 1M NaOH in excess (approx 75 mL) with constant stirring. Wash the blue precipitate twice with 100 mL of water in a Buchner funnel through two layers of Whatman 41 filter paper until the pH of the wash is <9. Partially dry on the funnel. Scrape off the precipitate and dissolve it in 40 mL of fresh 0.88 ammonia solution (deep blue) to form a saturated solution. Stir well for at least 15 min before use.

5. Borate buffer (0.2*M*) used when coupling protein to diazocellulose: 12.4 g of boric acid, 14.9 g of KCl/L adjusted to pH 8.2 with NaOH.
6. Anti-(IgG) immunoglobulin coupled to diazocellulose to make the immunoadsorbent may be an appropriate ammonium sulfate cut (between 40% and 50%) from serum or ascites, which is then well dialyzed against borate buffer solution.
7. Homogenizing solution for rat cerebral cortex: 11% sucrose (0.32*M*) containing 1 m*M* EDTA, 5 m*M* HEPES NaOH, 1 mg/mL bovine serum albumin, pH 7.4.
8. Krebs Ringer HEPES (KRH) solution: 140 m*M* NaCl, 5 m*M* KCl, 2.6 m*M* CaCl$_2$, 2 m*M* MgCl$_2$, 5 mM glucose, 10 m*M* HEPES NaOH, pH 7.4.
9. KRH + EDTA: KRH + 1 m*M* EDTA, pH 7.4, without Ca$^{2+}$ and Mg$^{2+}$.
10. 45% (v/v) Percoll™ in KRH + EDTA solution.
11. The first antibody used in the preparation of cholinergic nerve terminals is sheep anti-(Torpedo Chol-1), a polyclonal antiserum raised against Torpedo electric organ, which specifically reacts with a ganglioside uniquely found in cholinergic nerve terminals in many phyla (*18*). The second antibody, coupled to cellulose as immunoadsorbent, is a monoclonal mouse anti-(sheep IgG) antibody giving an immunoadsorbent bearing 0.3 mg mouse IgG/mg cellulose (*17*; *see* Note 1).
12. All other chemicals required are analytical grade and are acetic acid, benzene (**warning:** toxic, use in a fume cupboard), ethanol, HCl, H$_2$SO$_4$, β-naphthol, sodium acetate, sodium nitrite, and urea.
13. Apparatus required: appropriate centrifuge (Sorvall RC5) and rotors, glass Petri dish (9-cm diameter), oven (to 150°C), suitable Teflon-glass homogenizer with overhead drive (640 rpm), scintered glass funnel.

# 3. Methods

## 3.1. Aminocellulose Preparation

1. Dissolve 0.5 g of sodium acetate in 2 mL of water, and 1.4 g of m-nitrobenzyloxymethyl pyridinium chloride in 18 mL of ethanol.
2. Mix the two solutions, add 5 g of cellulose powder, and stir to a slurry.
3. Heat to 70°C in a glass Petri dish until dry (approx 30 min), and then for a further 40 min at 125°C.
4. Wash three times with 200 mL of benzene in a scintered glass funnel and suck dry. Wash with 1 L of water.
5. Resuspend in 150 mL of sodium dithionite, and mix at 55–60°C for 30 min.
6. Wash three times with 200 mL of water, twice with 200 mL of 30% acetic acid, and again with water to remove all traces of H$_2$S.

7. Dry in a desiccator at 20°C. If lumpy, powder the aminocellulose and store desiccated at 4°C, where it is stable for several months.

## 3.2. Diazocellulose Preparation

1. Dissolve 0.5 g of the powdered aminocellulose in 40 mL of ammonia-cal cupric hydroxide solution. Stir well for at least 15 min.
2. Centrifuge (approx 5000$g$ for 5 min) to remove any excess cupric hydroxide and/or any remaining lumps of aminocellulose.
3. Decant into 1500 mL of water to give a light blue solution. Add 1.8$M$ $H_2SO_4$ until the solution is colorless or very pale blue and the pH <4 when the aminocellulose flocculates to form a white precipitate.
4. Leave to stand for 30 min to settle, and then siphon off as much water as possible.
5. Wash the precipitate four times with cold water (100-mL aliquots), centrifuging between washes. If the cellulose becomes difficult to spin down, add a drop or two of 1$M$ HCl.
6. Cool at 4°C. The cellulose may turn pink.
7. Suspend the aminocellulose in 50 mL of cold fresh 2$M$ HCl. Add 2 mL of 1% (w/v) sodium nitrite, and test for free oxidant using starch iodide paper that turns black. Then mix for 20 min at 4°C.
8. Add excess solid urea (approx 5 g) until the starch iodide test is almost negative, and then wash at 4°C three times with water and twice with 0.2$M$ borate buffer. Centrifuge between washes to collect the diazocellulose.
9. Suspend in a small volume (25 mL) of 0.2$M$ borate buffer. To test for completeness of diazotization add some to a little β-naphthol in water—it should turn bright orange.

## 3.3. Protein-Cellulose Coupling and Immunoadsorbent Preparation

1. Immediately after preparing diazocellulose, add an equal quantity (usu-ally 100 mg/100 mg protein) to the mouse anti-(sheep IgG) protein in borate buffer (approx 15 mL total).
2. Mix for 48 h at 4°C in the dark.
3. Centrifuge and retain the supernatant, which can be used again for coupling.
4. Wash the immunoadsorbent three times in 0.2$M$ borate buffer.
5. Resuspend the immunoadsorbent at 5 mg/mL in 0.2$M$ borate buffer.
6. Measure the amount of protein bound to the immunoadsorbent by the method of Lowry et al. (*19; see also* Chapter 18). Usually 200–300 µg protein is bound/mg cellulose.

### 3.4. Immunoaffinity Purification
### of Cholinergic Nerve Terminals (see Notes 2–4)

1. Using 1 g of scraped rat cerebral cortex make a 10% (w/v) tissue homogenate in 11% sucrose solution by using 12 up-and-down strokes in a rotating (640 rpm) Teflon/glass (loose-fitting) homogenizer. Keep at 4°C.
2. Centrifuge at 1000*g* for 10 min, take the supernatant and centrifuge at 18,000*g* for 20 min. Adjust the pellet (P2, *see* Table 1) to 10 mL with the 11% sucrose solution.
3. Divide into two 5-mL portions. Incubate one with sheep anti-(Chol-1) serum (1:25 to 1:200 dilution), and the other with a similar dilution of nonimmune sheep serum, for 40 min at 4°C with periodic shaking. Dilute to 30 mL with 11% sucrose.
4. Centrifuge at 18,000*g* for 15 min (Sorvall SS-34, 14,000 rpm).
5. Wash the pellets at least three times with 30 mL of 11% sucrose, centrifuging at the aforementioned speed between washes.
6. Resuspend the pellets in 3 mL of Percoll solution, and centrifuge at 9000*g* for 2 min. Take the top layer that contains nerve terminals, dilute with KRH + EDTA solution, and centrifuge again.
7. Incubate four 200-µL aliquots of each pellet with 1–2 mg of mouse anti-(sheep IgG) immunoadsorbent in a total volume of 400 µL. Ensure good mixing. Incubate for 40 min at 4°C with occasional swirling.
8. Separate the bound and free nerve terminals by the addition of 2.5 mL of Percoll solution, mix and centrifuge at 9000*g* for 2 min.
9. Remove the supernatant. Wash the immunoadsorbent three times with 2.5 mL of Percoll solution and once with KRH solution.

## 4. Notes

1. With the indirect technique of immunoaffinity purification for organelles and membrane fractions, both the choice of first antibody and immunoadsorbent are of great importance. The first antibody may be an affinity purified polyclonal antibody or a monoclonal antibody directed at an abundant membrane antigen. Alternatively, it may be a crude antiserum from which specific antibodies are effectively affinity-purified by adding to the membrane preparation at 0°C (to prevent vesiculation and processes such as endocytosis), followed by subsequent washing. Antibodies against added antigenic determinants have also been used after either haptenization of the membranes or incorporation of viral coat proteins.

Table 1
Affinity Purification of Cholinergic Nerve Terminals from Rat Cerebral Cortex

| | P2 content | Nonspecific binding to adsorbent | Purified terminals on immunoadsorbent | Purification (X) | Yield, (%) |
|---|---|---|---|---|---|
| Choline acetyltransferase | 165 | 0.5 | 15.6 | 18.9 | 9.5 |
| Lactate dehydrogenase | 145,000 | 100 | 1100 | 1.5 | 0.8 |
| Glutamate decarboxylase | 474 | <0.8 | 1.1 | 0.5 | 0.2 |
| DOPA decarboxylase | 240 | 0.1 | 0.1 | 0.8 | 0.4 |
| Cytochrome oxidase | 11,800 | 100 | 400 | 5.9 | 3.3 |
| Acetylcholinesterase | 9900 | 100 | 700 | 14.1 | 7.1 |
| 2':3'-Cyclic nucleotide phosphodiesterase | 327,000 | 1300 | 2400 | 1.5 | 0.7 |
| 5'-Nucleotidase | 995 | <2.0 | 4.2 | 0.8 | 0.4 |

Membrane fractions (P2) from rat cortex were incubated with sheep anti-(Chol-1) serum or nonimmune serum (for nonspecific binding), washed, and incubated with immunoadsorbent. The immunoadsorbent pellets were assayed for various enzymes as listed. All data in the table is from at least six preparations expressed as nmol/h as previously described (17). Purification is expressed as the ratio of specific activity in the purified terminals to that in the P2 fraction. Note that the specific marker choline acetyltransferase was purified 18.9-fold over the P2 fraction, which is equivalent to a 38-fold purification from the original tissue. A 38-fold purification of the cholinergic terminals over glutamate decarboxylase was also achieved. No significant DOPA decarboxylase (a marker for terminals containing catecholamines) activity was found in the purified terminals. The major contaminant of the cortical preparation was the myelin enzyme 2':3'-cyclic nucleotide phosphodiesterase, which was purified 1.5-fold, with a yield of only 0.8%. Nonspecific binding to the immunoadsorbent after of all enzymes assayed was <1%. Electron microscopy showed a number of intact nerve terminals present on immunoadsorbent after affinity purification from the cortex. They contain synaptic vesicles and mitochondria as well as having postsynaptic adhesions.

The main requirement of the solid-phase immunoadsorbent is that it should have a high binding capacity. This depends both on the chemical nature of the support and the nature of the antibody-binding reagent (either an antiimmunoglobulin second antibody or Protein A from *Staphylococcus aureus*). It is possible to make a potentially "universal immunoadsorbent" bearing avidin and using a panel of biotinylated second antibodies. The high affinity of biotin for avidin ($K = 10^{15}\ M^{-1}$) permits very tight binding, whereas the ability of avidin to bind four biotin molecules potentially increases the capacity of the immunoadsorbent. The use of Staphylococcal Protein A on the solid support is limited by the fact that it binds to some species of immunoglobulin better than others and is especially poor at binding sheep and mouse IgG.

2. The easiest and most informative way of assessing purification is to calculate the percent yield of a specific component and per percent yield of nonspecific component (*see* Table 1 for cholinergic nerve terminals). There are, however, two ways of assessing the nonspecific component: One is the nonspecific binding of organelles to the solid support, either in the absence of the primary antibody or with a nonspecific immunoadsorbent, and the second is the amount of a given contaminant that copurifies with the desired membrane. Under some circumstances, the latter can also give an indication of the specificity of the primary antibody.

   Alternatively, purification can be expressed by the relative specific activity of a membrane component, i.e., the ratio of its specific activity in the purified membrane to that in the homogenate. One problem with this indicator is that when using antibody-coated immunoadsorbents it may be difficult to obtain accurate values for the protein content of the purified fraction.

3. It is usually difficult to separate membrane fractions from the immunoadsorbent without using extreme conditions, but elution is rarely necessary since most immunoaffinity-purified particles retain their normal biochemical functions while attached to the solid support. Thus the affinity-purified cholinergic synaptosome preparation is osmotically sensitive and metabolically active while still attached to the solid support *(17,20)*.

4. The major limitation on the more widespread use of immunoisolation to purify subcellular organelles and membranes is the availability of suitable first antibodies. This is particularly true of intracellular organelles since it has proven particularly difficult to raise antibodies against cytoplasmic exposed epitopes using subcellular fractions as antigens *(11)*. An attractive, potential approach to overcoming this problem is the use, as anti-

gens, of bacterially expressed proteins or synthetic peptides encoding membrane protein cytoplasmic domains predicted on the basis of molecular cloning. Cellulose or magnetic impermeable bead immunoadsorbents are suitable for most organelle and membrane immunoisolation, though new developments are likely to include an immunoadsorbent combining the flexibility of cellulose fibers and the ease of magnetic separation while avoiding nonspecific entrapment in a fibrous support.

## References

1. Luzio, J. P., Newby, A. C., and Hales, C. N. (1974) Immunological isolation of rat fat cell plasma membranes. *Biochem. Soc. Trans* **2**, 1385–1386.
2. Luzio, J. P., Newby, A. C., and Hales, C. N. (1976) A rapid immunological procedure for the isolation of hormonally sensitive rat fat cell plasma membrane. *Biochem. J.* **154**, 11–21.
3. Ito, A. and Palade, G. E. (1978) Presence of NADPH-cytochrome P-450 reductase in rat liver Golgi membranes. Evidence obtained by immunoadsorption method. *J. Cell Biol.* **79**, 590–597.
4. Richardson, P. J. and Luzio, J. P. (1986) Immunoaffinity purification of subcellular particles and organelles. *Appl. Biochem. Biotechnol.* **13**, 133–145.
5. Howell, K. E., Ansorge, W., and Gruenberg, J. (1988) Magnetic free-flow immunoisolation system designed for subcellular fractionation, in *Microspheres: Medical and Biological Applications* (Rembaum, A. and Tokes, Z. A., eds.), CRC, Boca Raton, FL, pp. 33–52.
6. Howell, K. E., Gruenberg, J., Ito, A., and Palade, G. E. (1988) Immunoisolation of subcellular components, in *Cell-Free Analysis of Membrane Traffic* (Morre, D. J., Howell, K. E., Cook, G. M. W., and Evans, W. H., eds.), Liss, New York, pp. 77–90.
7. Hubbard, A. L., Dunn, W. A., Mueller, S. C., and Bartles, J. R. (1988) Immunoisolation of hepatocyte membrane compartments using S. Aureus cells, in *Cell-Free Analysis of Membrane Traffic* (Morre, D. J., Howell, K. E., Cook, G. M. W., and Evans, W. H., eds.), Liss, New York, pp. 115–127.
8. Lea, T., Vartdal, F., Nustad, K., Funderud, S., Berge, A., Ellingsen, T., Schmid, R., Stenstad, P., and Ugelstad, J. (1988) Monosized, magnetic polymer particles: Their use in separation of cells and subcellular components, and in the study of lymphocyte function *in vitro. J. Mol. Recog.* **1**, 9–18.
9. Luzio, J. P., Mullock, B. M., Branch, W. J., and Richardson, P. J. (1988) Immunoaffinity techniques for the purification and functional assessment of subcellular organelles, in *Cell-Free Analysis of Membrane Traffic* (Morre, D. J., Howell, K. E., Cook, G. M. W., and Evans, W. H., eds.), Liss, New York, pp. 91–100.
10. Richardson, P. J. and Luzio, J. P. (1988) Immunoaffinity purification of membrane fractions from mammalian cells, in *Subcellular Biochemistry*, vol. 12 (Harris, J. R., ed.), Plenum, New York, pp. 221–241.

11. Howell, K. E., Schmid, R., Ugelstad, J., and Gruenberg, J. (1989) Immunoisolation using magnetic solid supports: subcellular fractionation for cell-free functional studies. *Methods Cell Biol.* **31,** 265–292.
12. Devaney, E. and Howell, K. E. (1985) Immunoisolation of a plasma membrane fraction from the Fao cell. *EMBO J.* **4,** 3123–3130.
13. Gruenberg, J. and Howell, K. E. (1985) Immunoisolation of vesicles using antigenic sites either located on the cytoplasmic or exoplasmic domain of an implanted viral protein. A quantitative analysis. *Eur. J. Cell Biol.* **38,** 312–321.
14. Hales, C. N. and Woodhead, J. S. (1980) Labelled antibodies and their use in the immunoradiometric assay. *Methods Enzymol.* **70,** 334–355.
15. de Curtis, I. and Simons, K. (1989) Isolation of exocytic carrier vesicles from BHK cells. *Cell* **58,** 719–727.
16. Wandinger-Ness, A., Bennett, M. K., Antony, C., and Simons, K. (1990) Distinct transport vesicles mediate the delivery of plasma membrane proteins to the apical and basolateral domains of MDCK cells. *J. Cell Biol.* **111,** 987–1000.
17. Richardson, P. J., Siddle, K., and Luzio, J. P. (1984) Immunoaffinity purification of intact, metabolically active, cholinergic nerve terminals from mammalian brain. *Biochem. J.* **219,** 647–654.
18. Richardson, P. J., Walker, J. H., Jones, R. T., and Whittaker, V. P. (1982) Identification of a cholinergic-specific antigen Chol-1 as a ganglioside. *J. Neurochem.* **38,** 1605–1614.
19. Lowry, O. H., Rosebrough, N. J., Farr, A. L., and Randall, R. J. (1951) Protein measurement with the Folin phenol reagent. *J. Biol. Chem.* **193,** 265–275.
20. Richardson, P. J., Brown, S. J., Bailyes, E. M., and Luzio, J. P. (1987) Ecto-enzymes control adenosine modulation of immunoisolated cholinergic synapses. *Nature* **327,** 232–234.

CHAPTER 14

# Separation and Analysis of Phospholipids by Thin Layer Chromatography

## Ian J. Cartwright

## 1. Introduction

The membranes of animal cells contain a substantial amount of lipid, which may account for between 20 and 80% of the membrane mass. The predominant lipid components are phospholipids (PL) and cholesterol; with glycolipids, cholesterol esters, glycerides, and free fatty acids as minor constituents. PL perform a basic structural role in the cell membrane. They are arranged in a bilayer configuration, in which proteins are embedded, or associated peripherally. However, PL do not merely form an inert framework in membranes, they also play important functional roles in intracellular signaling, secretion, membrane transport, endocytosis, and cell fusion.

In order to analyze cellular PL composition, it is first necessary to employ an extraction procedure that quantitatively recovers total lipids in a relatively pure state. The effectiveness of lipid extraction procedures largely depends on the chemical nature of the lipid components and how they associate with each other and with membrane proteins. The main forces that bind lipids to membrane proteins are hydrogen bonding and electrostatic and hydrophobic association. Thus, in order to liberate membrane lipids, it is necessary to disrupt the membrane structure and denature the membrane proteins with a polar solvent, such as methanol or ethanol. The methods most commonly used

From: *Methods in Molecular Biology, Vol. 19: Biomembrane Protocols: I. Isolation and Analysis*
Edited by: J. M. Graham and J. A. Higgins  Copyright ©1993 Humana Press Inc., Totowa, NJ

to extract membrane lipids are based on a procedure originally described by Folch et al. *(1)*, and employ chloroform and methanol.

If subsequent analysis of the fatty acid (FA) components of membrane lipids is required, it is important to add an antioxidant to the extraction solvents, in order to minimize oxidative damage to unsaturated FA. After the initial, crude lipid extract has been obtained, it must be subjected to a washing step, in order to remove any nonlipid contaminants, such as free sugars and amino acids. The purified lipid extract is then dried, redissolved in a small volume of solvent, and stored in an inert atmosphere, at –70°C, until required for further analysis by thin-layer chromatography (TLC).

TLC is a highly effective and versatile technique for separation of intact PL, as well as for glycolipids (*see* Chapter 26) and neutral lipids *(2,3)*. Lipids are usually separated on layers of silica gel adsorbent held on glass plates. The lipid samples are applied to plates as discrete spots or streaks and developed in a solvent mixture. The solvents carry the various lipid components up the plate at different rates, according to the extent to which they are held by the adsorbent. The advantages of TLC are numerous, including simplicity and ease of manipulation, rapidity, sensitivity, and high resolving power. Simple lipid mixtures can normally be separated by TLC in one dimension (1D-TLC). For complex lipid mixtures, the resolution can be greatly increased by development in two dimensions (2D-TLC). After the lipid classes have been resolved, they can be identified on the plate, using various detection reagents. The areas of silica containing the adsorbed lipids can then be isolated, scraped from the plate and taken for further analysis, including quantitation of FA components by gas-liquid chromatography (*see* Chapter 17) and PL phosphorus assay.

The aim of this chapter is to describe the basic methods for the analysis of membrane PL composition; including procedures for the extraction and purification of lipids, and for the quantitation of PL components after separation by TLC.

## 2. Materials
### 2.1. Glassware

1. Disposable glass Pasteur pipets.
2. Pyrex glass tubes (6-mL capacity), with Teflon-lined screw-caps; for storage of purified total lipid extracts, prior to TLC separation. After

lipid separation, these tubes can be used for storage of samples of individual lipid components bound to TLC adsorbents.

3. Pyrex glass centrifuge tubes (20-mL capacity) with ground-glass stoppers; for use in lipid extraction procedures.
4. Pyrex glass test tubes (10-mL capacity); for use in PL phosphorus assays.

## 2.2. Apparatus

1. A Potter-Elvehjem tissue homogenizer.
2. Long-tipped, calibrated, positive-displacement pipets (0.1-, 0.2-, 1.0-, and 5-mL capacities), for dispensing solvents and lipid solutions.
3. A supply of nitrogen gas (oxygen-free) for evaporation of small volumes of solvents (<20 mL) from lipid extracts and for drying of TLC plates.
4. A thermostatically controlled, aluminum heating-block (with holes drilled to accommodate 6-, 10-, and 20-mL test tubes) for incubation of lipid extracts during solvent evaporation under a nitrogen stream.
5. A rotary evaporator; for removal of large volumes of solvents (>20 mL) from lipid extracts. This apparatus should have greaseless joints, to prevent contamination of lipid samples.
6. A microsyringe (10-μL capacity) for application of lipid samples to TLC plates.
7. Glass TLC plates (20 × 20 cm) precoated with silica gel H at a thickness of 0.25 mm. These are commercially available from a number of sources (e.g., Merck [Darmstadt, Germany], Analab [North Haven, CT]).
8. Rectangular, glass TLC tanks (11 × 21 × 21 cm; depth × width × height) with ground glass covers.
9. A plate drying box; for evaporation of solvents from TLC plates in an atmosphere of nitrogen. This consists of a perspex box (25 × 45 × 3 cm; depth × width × height) with one open side (25 × 45 cm), an inlet for nitrogen, and an outlet for solvent vapors. One or two TLC plates are placed inside the box (which is mounted on a hot plate), heated to 30°C, and dried under the stream of nitrogen. The whole assembly should be positioned in a fume cupboard. A drying box can be cheaply and easily manufactured in any laboratory workshop.
10. A spray-bottle, powered by an aerosol can. This is used to spray reagents onto TLC plates, in order to detect the lipid bands. Spray-bottles can be obtained from suppliers of TLC plates.
11. A UV light source, either long-wave (265 nm) or short-wave (254 nm); for visualization of lipid bands on TLC plates sprayed with fluorescent indicator.
12. A spectrophotometer.

## 2.3. Reagents

1. 30% aq. ammonia.
2. $CaCl_2$ (anhydrous).
3. Potassium oxalate (analytical grade).
4. Antioxidant: butylated hydroxytoluene (BHT). Add 150 mg/L to all organic solvents employed in extraction and TLC of lipids. This minimizes oxidative damage of unsaturated FA components, which is especially important if there is a requirement for their quantitative analysis by gas-liquid chromatography (GLC; *see* Chapter 17).
5. Organic solvents: Glacial acetic acid, acetone, chloroform, diethyl ether, n-heptane, n-hexane, methanol, petroleum ether. These should all be analytical grade (*see* Notes 1 and 2).
6. Lipid detection reagents: bismuth nitrate (pentahydrate), 1,6-diphenyl-1,3,5-hexatriene (DPH), iodine, mercury, 1-naphthol, ninhydrin, KI (analytical grade), conc. $H_2SO_4$ (*see* Section 3.4.).
7. Phosphorus assay reagents: Fiske-Subarrow Reducer (Sigma Chemical Co., St. Louis, MO), ammonium molybdate (tetrahydrate), conc. HCl, perchloric acid, $Na_2HPO_4 \cdot 12H_2O$. The components of the commercial reducer can be purchased separately (*see* Section 12. of Chapter 1).

## 2.4. Lipid Standards

1. Phospholipids (PL): Phosphatidic acid (PA), phosphatidylglycerol (PG), diphosphatidylglycerol (DPG), phosphatidylcholine (PC), phosphatidylethanolamine (PE), phosphatidylinositol (PI), phosphatidylinositol monophosphate (PI-P), phosphatidylinositol diphosphate (PI-PP), phosphatidylserine (PS), sphingomyelin (SM), lysophosphatidylcholine (LPC), lysophosphatidylethanolamine (LPE), lysophosphatidylinositol (LPI), lysophosphatidylserine (LPS). Dissolve 1 mg of each standard in 1 mL of chloroform/methanol (1:1 [v/v]), and store under nitrogen at –20°C in a 6-mL Pyrex tube, with a Teflon-lined, screw-cap.
2. Glycolipids (GL): ceramide monohexosides (CM), gangliosides (GS). Prepare and store standard solutions as in item 1.
3. Neutral lipids (NL): Cholesterol (C), cholesterol ester (CE), monoglyceride (MG), diglyceride (DG), triglyceride (TG), fatty acid (FA; e.g., stearic acid). Prepare and store standard solutions as in item 1.

## 3. Methods

### 3.1. Lipid Extraction from Tissues or Concentrated Membrane Suspensions

1. For whole tissue (e.g., rat liver) extraction, use 0.5 g of fresh tissue homogenized in 2 mL of water. For membrane extraction, use

2 mL of aqueous suspension, containing up to 100 mg (dry wt) of membranes.

2. Transfer the tissue homogenate or membrane suspension to a 20-mL centrifuge tube, containing a magnetic stir-bar. Add 10 mL of chloroform/methanol (2:1 [v/v]) and stir for 5 min.

3. Centrifuge the mixture at 620g for 10 min, transfer the supernatant to a conical flask, and maintain on ice.

4. Re-extract and centrifuge the residue as above, and combine the supernatants.

5. Re-extract the residue with 10 mL of chloroform/methanol (2:1 [v/v]); containing conc. HCl (0.2 mL/L), in order to ensure quantitative recovery of the phosphoinositides (i.e., PI, PI-P, and PI-PP).

6. Recentrifuge the residue, pool the supernatants, and evaporate the lipids to dryness in a rotary evaporator (*see* Notes 3 and 4).

7. Redissolve the dry, crude lipid extract in 4 mL of chloroform/methanol (2:1 [v/v]) and transfer to a preweighed, 6-mL Pyrex tube, with a Teflon-lined screw-cap.

8. Wash the lipid extract by the addition of 0.8 mL of 50 m$M$ CaCl$_2$, followed by vigorous mixing for 1 min. Allow the mixture to stand until it separates into two distinct phases. The lower organic phase contains the lipids, with the exception of GS, which remains in the upper aqueous phase, together with the nonlipid contaminants.

9. Using a Pasteur pipet, carefully transfer the upper phase to a separate test tube. If required, GS can be isolated from this phase by dialysis and lyophilization.

10. Transfer the tube containing the organic phase to a heating block, and evaporate the lipids to dryness under a nitrogen stream, at 30°C.

11. Reweigh the tube and determine the dry wt of lipid.

12. Redissolve the dry lipid extract in 200 µL of chloroform, flush out the air in the tube with nitrogen, reseal and store at –70°C, prior to further analysis by TLC or GLC (*see* Chapter 17).

### *3.2. Lipid Extraction from Dilute Membrane Suspensions*

Large volumes of dilute suspensions of membranes in aqueous media are best extracted by the method of Bligh and Dyer *(4)*, as follows.

1. To each 10 mL of membrane suspension, add 40 vol of chloroform/methanol (1:2 [v/v]) and mix to produce a single phase. If large precipitates form, these should be removed by filtration or centrifugation. If precipitates are small they can be removed later, following phase separation.

2. Add 12 mL of chloroform and 12 mL of water, mix and allow the two phases to separate.

3. Using a Pasteur pipet, remove the upper phase (containing most of the nonlipid contaminants), together with any denatured protein at the interface. The lower phase contains the lipids, with the exception of GS, which can be recovered from the upper phase, as described earlier for whole tissue.

4. Evaporate the lipids to dryness in a rotary evaporator, redissolve in 200 μL of chloroform, and store exactly as above for whole tissue.

## 3.3. Separation of PL by TLC

The basic techniques for the separation of PL by 1D-TLC and 2D-TLC are described in the following sections. The list of solvent systems presented here for the resolution of lipids is by no means exhaustive, and the reader should consult the more extensive texts for other alternatives *(2,3)*.

### 3.3.1. Application of Lipid Samples and Development of TLC Plates (see Notes 5 and 6)

1. Line the TLC tank with filter paper and add the appropriate solvent mixture (freshly prepared) (*see* Section 3.3.2.) to a depth of 1 cm. Seal the tank lid with petroleum jelly, and allow the solvents to saturate the atmosphere inside the tank. This should be done at least 1 h before development of plates. A maximum of two plates should be run in each tank.

2. Evaporate the lipid samples to dryness under a stream of nitrogen, and redissolve in chloroform/methanol (1:1 [v/v]) to give a 5–10% solution.

3. Activate the TLC plate by heating at 100°C for 30 min, and allow to cool.

4. For 1D-TLC, use a soft pencil to divide the plate into five columns (i.e., four sample lanes and one standard lane), taking care not to damage the adsorbent layer. Using a microsyringe, apply 1–5 mg of each lipid sample as a streak (0.5- to 1.0-cm wide), at 2.5 cm from the bottom edge of the plate. Also apply 10 μg of standard (either one or a mixture) as a single spot. For 2D-TLC, apply only one lipid sample (1–5 mg) per plate, as a single spot in the bottom right-hand corner, at about 2.5 cm from the bottom and side edge of the plate. A standard can also be applied adjacent to the sample, at 1 cm from the side edge.

5. Evaporate the solvents from the lipid streaks or spots under a stream of nitrogen.

6. Immediately develop the plate in a tank containing the appropriate solvent system (*see* Sections 3.3.2. and 3.3.3.). Allow the solvents to ascend

to about 1 cm from the top of the plate, remove the plate, and lightly mark the solvent front with a soft pencil (*see* Notes 5 and 6).

7. Dry 1D-TLC plates in air or under nitrogen (in a plate-drying box), if subsequent GLC analysis of FA components is required. Dry 2D-TLC plates under nitrogen for 20 min, turn clockwise through 90°C, and redevelop in the second solvent system. Allow the solvents to ascend to 1 cm from the top of the plates. Remove the plates and dry in air or nitrogen, as required.

8. Spray the developed 1D/2D-TLC plates with the appropriate lipid detection reagent (*see* Section 3.4.).

### 3.3.2. Solvent Systems for 1D-TLC (see Fig. 1)

1. Separation of total PL: Plates are developed in hexane/diethyl ether/ glacial acetic acid (90:10:1 [v/v/v]). Total PL remain at the origin and NL separate as shown in Fig. 1 (Plate 1).

2. Separation of major PL classes: Plates are developed in chloroform/ methanol/glacial acetic acid/water (60:50:1:4 [v/v/v/v]). This system resolves DPG, PC, PE, PI, PS, and SM, as shown in Fig. 1 (Plate 2). Lyso-PL remain at the origin, and are not completely separated from each other. NL move with the solvent front.

3. Separation of major acidic PL: Plates are first developed in hexane/ acetone (3:1 [v/v]), dried under nitrogen, and redeveloped in the same direction in chloroform/methanol/glacial acetic acid/water (80:13:8:0.3 [v/v/v/v]). The first solvent system removes NL to the top of the plate, so that they do not overlap the acidic PL. The second solvent system resolves the acidic PL (PA, PG, DPG) and also PE and CM, as shown in Fig. 1 (Plate 3).

4. Separation of polyphosphoinositides: Prior to development silica gel H plates should be immersed for about 1 h in a solution of 1% potassium oxalate, which chelates calcium ions and prevents retardation in the movement of PI-PP (5). After drying (at 100°C) and sample application, the plates are developed in chloroform/methanol/30% aq. ammonia/water (48:40:5:10 [v/v/v/v]). This system separates PI-P and PI-PP from other PL, as shown in Fig. 1 (Plate 4).

### 3.3.3. Solvent Systems for 2D-TLC

Complex mixtures of lyso-PL and PL can be separated by the following 2D-TLC system. The plates are developed in the first dimension: chloroform/methanol/30% aq. ammonia/water (90:54:5.5:5.5 [v/v/v/v]) and dried under nitrogen. They are then rotated through 90° and developed in the second dimension: Chloroform/methanol/acetone/glacial

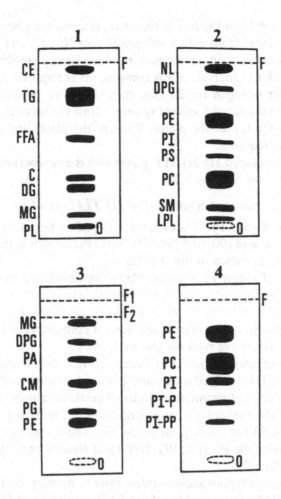

Fig. 1. Schematic 1D-TLC separations of lipids on silica gel H. Solvent systems: Plate 1, development with hexane/diethyl ether/glacial acetic acid (90:10:1 [v/v/v]) to F; Plate 2, chloroform/methanol/glacial acetic acid/water (60:50:1:4: [v/v/v/v]) to F; Plate 3, hexane/acetone (3:1 [v/v]) to $F_1$, followed by chloroform/methanol/glacial acetic acid/water (80:13:8:0.3 [v/v/v/v]) to $F_2$; Plate 4, methanol/chloroform/30% aq. ammonia/water (48:40:5:10 [v/v/v/v]) to F. Abbreviations: O, origin; NL, neutral lipid; TG, triglyceride; DG, diglyceride; MG, monoglyceride; C, cholesterol; CE, cholesterol ester; FFA, free fatty acid; PL, phospholipid; LPL, lysophospholipid; PG, phosphatidylglycerol; DPG, diphosphatidylglycerol; PA, phosphatidic acid; PC, phosphatidylcholine; PE, phosphatidylethanolamine; PI, phosphatidylinositol; PI-P, phosphatidylinositol monophosphate; PI-PP, phosphatidylinositol diphosphate; PS, phosphatidylserine; SM, sphingomyelin; CM, ceramide monohexosides.

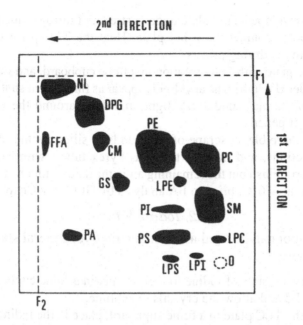

Fig. 2. Schematic 2D-TLC separation of lipids on silica gel H. Solvent systems: 1st dimension, development with chloroform/methanol/30% aq. ammonia/water (90:54:5.5:5.5 [v/v/v/v]); 2nd dimension, development with chloroform/methanol/acetone/glacial acetic acid/water (60:20:80:20:10 [v/v/v/v/v]). Abbreviations: LPC, lysophosphatidylcholine; LPE, lysophosphatidylethanolamine; LPI, lysophosphatidylinositol; LPS, lysophosphatidylserine; GS, gangliosides. For other abbreviations, *see* legend to Fig. 1.

acetic acid/water (60:20:80:20:10 [v/v/v/v/v]). These systems resolve all the individual PL and lyso-PL, and also CM and GS, as depicted in Fig. 2.

## 3.4. Detection of Lipids on TLC Plates (see Note 7)

### 3.4.1. Diphenyl Hexatriene (DPH)

All acyl lipids can be detected with the nondestructive fluorescent probe, 1,6-diphenyl-1,3,5-hexatriene *(6)*, using the following method. The reagent does not interfere with the subsequent GLC analysis of FA components.

1. Dissolve 10 mg of DPH in 100 mL of petroleum ether, and store the solution in a dark glass bottle. The reagent is stable at room temperature for about 2 mo.

2. Dry the developed TLC plate under a stream of nitrogen inside a plate-drying box, mounted on a hot-plate. Heat the TLC plate to 30°C to accelerate the drying process.
3. Spray the plate with DPH solution in a fume cupboard, and immediately view under UV light. The acyl lipids appear as fluorescent spots or bands.
4. Maintain the plate under UV light, and draw around the lipid bands with a soft pencil.
5. Using a sharp blade, scrape off the bands of silica containing the acyl lipid fractions, and transfer to 6-mL Pyrex tubes, with Teflon-lined screw-caps. Flush out the remaining air in the tubes with nitrogen, reseal, and store at −70°C, prior to FA analysis by GLC (*see* Chapter 17).

### 3.4.2. Iodine Vapor

Iodine vapor can be used as a nondestructive, general stain for lipids, as follows.

1. Add a few crystals of iodine to a glass tank in a fume cupboard. Close the tank lid and allow the crystals to sublime.
2. Air-dry the TLC plate in a fume cupboard, place in the iodine tank, and close the lid. The lipids appear as brown to yellow bands or spots. The color development time is dependent on the amount of lipid on the plate, and may be minutes to hours.
3. Draw around the lipid bands with a soft pencil, and allow the iodine to fade in a fume cupboard; this can be accelerated by warming the plate. If lipids are to be determined colorimetrically, a residual trace of iodine in the silica bands should not interfere with the assay.

### 3.4.3. Phosphate Stain

All PL can be detected by the extremely sensitive method of Vaskovsky and Kostevsky (7), as follows.

1. Prepare the following solutions (*see* Note 8):
   Solution A:  Dissolve 16 g of ammonium molybdate in 120 mL of distilled water.
   Solution B:  To 40 mL of conc. HCl, add 10 mL of mercury and 80 mL of Solution A. Shake the mixture for 30 min and filter.
   Solution C:  To 200 mL of conc. $H_2SO_4$, add all of Solution B and the remainder of Solution A, and dilute to 1 L with distilled water. This solution is stable indefinitely if stored in a dark glass bottle.
2. Air-dry the developed TLC plate in a fume cupboard, and then spray with Solution C. After a few minutes, PL appear as dark blue bands or spots. The stain does not interfere with subsequent colorimetric assay of PL.

### 3.4.4. Ninhydrin Stain

The amino groups of PE and PS can be detected specifically with ninhydrin, as follows.

1. Air-dry the developed TLC plate in a fume cupboard, and then spray with ninhydrin solution (2.5 g/L in acetone).
2. Heat the plates at 100°C, and the amino-PL appear as pink purple bands or spots within 15 min.

### 3.4.5. Choline Stain

The choline-containing PL (i.e., PC and SM) can be specifically detected with the Dragendorff stain *(8)*, as follows.

1. Prepare the following solutions:
   Solution A: Dissolve 1.7 g of bismuth nitrate in 100 mL of acetic acid (20% aq.).
   Solution B: Dissolve 10 g of KI in 25 mL of distilled water.
   Solution C: Mix 20 mL of Solution A and 5 mL of Solution B, and dilute to 70 mL with distilled water.
2. Air-dry the developed TLC plate in a fume cupboard, and then spray with Solution C. After a few minutes, PC and SM appear as orange bands or spots.

### 3.4.6. 1-Naphthol Stain

Glycolipids can be specifically detected with 1-naphthol stain *(9)*, as follows.

1. Prepare the following solutions:
   Solution A: Dissolve 0.5 g of 1-naphthol in 100 mL of methanol/water (1:1 [v/v]).
   Solution B: Add 95 mL of conc. $H_2SO_4$ to 5 mL of distilled water.
2. Air-dry the developed TLC plate in a fume cupboard, and then spray with Solution A.
3. Air-dry the plate again, and respray with Solution B.
4. Heat the plate at 120°C until the color develops. Glycolipids stain purple to blue, whereas other polar lipids stain yellow.

## 3.5. Elution of Lipids from TLC Adsorbents

1. After detection by a suitable nondestructive stain, score around each lipid band on the TLC plate with a needle.
2. Using a sharp blade, scrape off each band of silica-bound lipid and transfer it to a 15-mL glass centrifuge tube.

3. For elution of nonacidic lipids (i.e., C, CE, MG, DG, TG, LPC, PC, LPE, PE, SM, CM, GS), use chloroform/methanol (1:1 [v/v]). For elution of acidic lipids (i.e., FFA, PA, DPG, PG, LPI, PI, PI-P, PI-PP, LPS, PS), it is necessary to acidify the chloroform/methanol (1:1 [v/v]) with conc. HCl (0.2 mL/L). Add 14 mL of the appropriate eluting solvent mixture to each tube, vortex-mix, and pellet the silica by centrifugation at 1000$g$ for 10 min.
4. Using a glass Pasteur pipet, transfer each supernatant to a 50-mL conical flask.
5. Repeat the extraction and centrifugation steps twice more for each sample.
6. Pool the supernatants and dry each lipid fraction in a rotary evaporator (*see* Notes 3 and 4).
7. Redissolve each dry lipid fraction in 1 mL of chloroform/methanol (1:1 [v/v]), and transfer to a 6-mL Pyrex tube, with screw-cap (Teflon-lined). Flush out the remaining air in the tube with nitrogen, reseal and store at –20°C, for further analysis.

## 3.6. Quantitation of PL

After separation by TLC, PL can be determined by phosphorus assay, which is sensitive to micromolar quantities. For determination of nanomolar quantities, it is necessary to employ scanning densitometry of PL after separation by high-performance TLC (HPTLC). PL can also be quantified by GLC of their FA components (*see* Chapter 17). The procedure for the PL-phosphorus assay is described in the following section, and is followed by a brief outline of the HPTLC/scanning densitometry method.

### 3.6.1. Assay of PL-Phosphorus

PL can be determined by phosphorus assay, either directly on the TLC adsorbent, or after prior elution, using the following procedure, which is based on the method of Fiske and Subbarow *(10)*, modified by Bartlett *(11)*.

1. Prepare the following assay reagents:
   a. Standard phosphate solution, $Na_2HPO_4 \cdot 12H_2O$ (3.58 g/L aq.). Add a few drops of chloroform to prevent bacterial growth.
   b. Perchloric acid (PCA; 70% aq.).
   c. Ammonium molybdate solution (25 g/L aq.).
   d. Fiske-Subbarow reducing solution; dissolve 3.97 g of the solid in 25 mL of distilled water, and filter into a dark glass bottle. The solution is stable for about 8 wk at 4°C.
2. Prior to assay, wash all glass test tubes and pipets in distilled water containing about 5 mL/L of conc. HCl, in order to remove all traces of

phosphate contamination. Rinse the tubes with several changes of distilled water.

3. For assay of PL bound to TLC adsorbent, transfer each band of silica gel to a 10-mL Pyrex test tube, and add 1.5 mL of PCA. For assay of PL in total lipid extracts and in fractions eluted from silica gel, transfer a known volume of each sample solution to a 10 mL Pyrex tube, evaporate the solvent under a nitrogen stream, and add 1.5 mL of PCA.

4. Prepare a reagent blank tube, containing 1.5 mL of PCA.

5. Cap each Pyrex tube with a marble, and heat to 230°C for twice the time required for each mixture to turn from brown to colorless or pale yellow (usually about 30 min). PL digestion must be performed in a heating block in a fume cupboard. Extreme care must be taken, as hot PCA may explode in the presence of organic material.

6. Allow all tubes to cool to room temperature, then add 2.5 mL of distilled water and vortex-mix.

7. Centrifuge the tube containing silica gel at 1000$g$ for 10 min, and transfer the supernatants to clean test tubes.

8. Prepare tubes for the standard curve, using 0–2.0 µmol of phosphate.

9. Add 1.5 mL of ammonium molybdate solution to all digested samples, reagent blank, and phosphate standards, and then vortex-mix the tubes.

10. Add 0.2 mL of reducing solution to all tubes and then vortex-mix.

11. Cap all tubes with marbles and heat them for 9 min in a boiling water bath, and then allow them to cool to room temperature.

12. Read the absorbance of the blue color in the phosphate standards at 830 nm, and plot the standard curve. Read the absorbance of the digested samples against the reagent blank, and determine the µmols of PL from the standard curve. This method is sensitive to <0.05 µmol and linear to 2.0 µmol of phosphate.

### 3.6.2. HPTLC and Scanning Densitometry of PL

When the lipid content of membrane samples is extremely low (nanomolar), the PL can be determined by *in situ* scanning densitometry, after separation by HPTLC. The HPTLC method employs layers of silica with a much finer grain size than on conventional TLC plates. This allows faster and more efficient separation of lipids and much lower detection limits (*see* Note 9).

1. For separation of acidic and nonacidic PL by HPTLC *(12)*, develop the plates first in chloroform/methanol/glacial acetic acid/formic acid/water (35:15:6:2:1 [v/v/v/v/v]); dry and then redevelop in hexane/diisopropyl ether/glacial acetic acid (65:35:2 [v/v/v]).

2. After development, dip the plates in a solution containing 3% cupric acetate (w/v) and 8% phosphoric acid (v/v), and char the PL by heating the plate to 140°C.
3. Quantify the PL using a scanning densitometer in the reflectance mode.

## 4. Notes

1. All solvents used in lipid extraction and TLC procedures should be handled in well ventilated areas, or in fume cupboards whenever possible.
2. There should be no naked flames in laboratories where flammable solvents are handled, and all electrical equipment should be correctly wired and earthed to reduce the risk of sparks.
3. Immediately prior to operation of a rotary evaporator, the equipment should be flushed out with nitrogen, in order to reduce the risk of autooxidation of unsaturated lipids.
4. A safety guard should be placed around the rotary evaporation apparatus when in operation, in order to protect the user against implosion of the flask.
5. It is important that the adsorbent layer on TLC plates should not touch the supporting side ridges in multiplate tanks, as this causes the ascending solvents to move more rapidly at the outer edges of the plate. This produces a curved solvent front, and the lipid bands run toward the center and overlap, rather than moving vertically up the plate.
6. Great care should be taken to avoid damaging the adsorbent layer on TLC plates, as this causes irregularity in the movement of solvent up the plate, and hence poor separation of lipids.
7. Before application of detection reagents, it is important that the TLC developing solvents are removed from the plates. If subsequent analysis of the FA components of acyl lipids is required, the plates must be dried under nitrogen, in order to minimize oxidative damage to unsaturated FA. Most of the lipid detection reagents are applied to TLC plates as a fine spray, which is most conveniently performed with a spray-bottle, powered by an aerosol can. To achieve an even coverage of detection reagent the TLC plate should be positioned at the bottom of a cardboard box (with the open side facing upward), and the spray applied in a fume cupboard.
8. The Vaskovsky and Kostevsky reagent is extremely toxic. This must be prepared with great care in a fume cupboard. Contact with the solution should be avoided. All operations with the reagent must be carried out in a fume cupboard. Consult your safety officer before commencement of this procedure.
9. Details of the methodology of HPTLC and scanning densitometry are outside the scope of this chapter; the reader should refer to Touchstone *(3)*.

# References

1. Folch, J., Lees, M., and Sloane-Stanley, G. H. (1957) A simple method for the isolation of total lipids from animal tissues. *J. Biol. Chem.* **226,** 497–509.
2. Christie, W. W. (1973) Chromatographic and spectroscopic analysis of lipids: General principles, in *Lipid Analysis*, Pergamon, Oxford, pp. 42–84.
3. Touchstone, J. C. (1986) Modern thin-layer chromatography. *Chem. Anal.* **85,** 149–178.
4. Bligh, E. G. and Dyer, W. J. (1959) A rapid method of total lipid extraction and purification. *Can. J. Biochem. Physiol.* **37,** 911–917.
5. Gonzalez-Sastre, F. and Folch-Pi, J. (1968) Thin-layer chromatography of the phosphoinositides. *J. Lipid Res.* **9,** 532–533.
6. Hoffman, W., Rink, D. A., Restall, C., and Chapman, D. (1981) Intrinsic molecules in fluid phospholipid bilayers: Fluorescence probe studies. *Eur. J. Biochem.* **114,** 585–589.
7. Vaskovsky, V. E. and Kostevsky, E. Y. (1968) Modified spray for the detection of phospholipids on thin-layer chromatograms. *J. Lipid Res.* **9,** 396.
8. Beiss, U. (1964) Separation of plant lipids by paper chromatography. *J. Chromatogr.* **13,** 104–110.
9. Kates, M. (1986) in *Techniques of Lipidology: Isolation, Analysis and Identification of Lipids* (Burdon, R. H. and Van Knippenberg, P. M., eds.), Elsevier Science, Amsterdam, p. 242.
10. Fiske, C. H. and Subbarow, Y. (1925) The colorimetric determination of phosphorus. *J. Biol. Chem.* **66,** 375–400.
11. Bartlett, G. R. (1959) Phosphorus assay in column chromatography. *J. Biol. Chem.* **234,** 466–468.
12. Wood, W. G., Cornwell, M., and Williamson, L. S. (1989) High performance thin-layer chromatography and densitometry of synaptic plasma membrane lipids. *J. Lipid Res.* **30,** 775–779.

CHAPTER 15

# Analysis of Phospholipids
# by High Performance
# Liquid Chromatography

*Hafez A. Ahmed*

## 1. Introduction
### 1.1. Background

Phospholipids are amphipathic molecules of great importance and widespread in biological material. They play an essential role in the structure and function of biological membranes. They are classified according to the nature of the "backbone" residue (glycerol or sphingosine), the type of nitrogenous base (or hexahydric alcohol) attached to it, and the nature of the chemical bonds (ester or ester and ether) linking the hydrocarbon chains to the backbone molecule. Within the same class, an enormous number of subclasses is possible through variation in the number of carbon atoms in the hydrocarbon chains, their degree of unsaturation, and, in the case of the glycerophospholipids, the position (C1 or C2) of the particular fatty acid (or ether) residues on the glycerol backbone. High performance liquid chromatography (HPLC) has emerged as one of the most powerful and versatile forms of separation technique that can be applied to the efficient separation and determination of phospholipids.

The actual step-by-step operation of the HPLC hardware is beyond the scope of this chapter and will depend on the particular model (*see* Note 1). Included instead is an Operational Strategy section (Section 1.2.), which lists some of the more important points regarding the

From: *Methods in Molecular Biology, Vol. 19: Biomembrane Protocols: I. Isolation and Analysis*
Edited by: J. M. Graham and J. A. Higgins Copyright ©1993 Humana Press Inc., Totowa, NJ

hardware and the choice of separation and detection systems. Sections 2. and 3. then give an example of a typical separation.

## 1.2. Operational Strategy

### 1.2.1. Use of Normal Phase vs Reverse Phase (1,2)

Phospholipid classes can be separated by normal-phase HPLC, where the mobile phase is hydrophobic (nonpolar) relative to the hydrophilic (polar) stationary phase. The method separates phospholipids according to their overall polarity. The stationary phase should be a good solvent for the sample but a poor solvent for the mobile phase. Now available are stable bonded packings, which are much more resistant to pressure and temperature fluctuations compared to the conventional coated phases.

In reverse-phase HPLC (3), the more nonpolar molecules selectively interact with the hydrophobic stationary phase and are retained more strongly than the more polar ones. Bonded phase supports provide a suitable stationary phase: The most popular bonded phase comprises a linear hydrocarbon, most commonly an octadecyl (C18) alkyl chain, but other chain lengths are also available. This technique can separate molecular species of a single phospholipid class according to the number of carbon atoms and the degree and type of unsaturation in the fatty acid chains. The separation of the molecular species of a phospholipid parallels the *equivalent carbon atom number*, which is defined as the total number of acyl chain carbon atoms minus the number of double bonds.

### 1.2.2. Separation Columns

Stainless steel straight columns of 25–50 cm in length and 1–4 mm in diameter, which can withstand pressures of approx $5.5 \times 10^7$ Pa are suitable. Silica or alumina are available as packing material with a good range of particle sizes. They pack efficiently and give good flow properties. In liquid-liquid partition systems the stationary phase is coated on to the inert support. All columns and transmission tubing must be protected from drafts and temperature fluctuations.

### 1.2.3. Guard Columns and In-Line Filters (1)

To prolong the life of analytical columns, guard columns, which act as physical and chemical filters, must be inserted ahead of the analytical columns. These are relatively short and contain a stationary phase similar to that in the analytical column. They protect the analytical columns from particulates that might either be present in a contaminated mobile phase or released from sample injection valves, when they have been in use for long periods. A guard column extends the lifetime of the separation column by capturing any strongly immobilized impurities from the sample and preventing them from contaminating the upper regions of the analytical columns. Guard columns must be periodically replaced, reconditioned, or repacked.

Guard columns can cause broadening of solute peaks on the analytical column with possible loss of resolution as a consequence. If this effect cannot be tolerated, an in-line filter that is designed to remove only particulates should be placed in front of the separation column. Details of suitable devices can be obtained from the manufacturers.

### 1.2.4. Pump System

Because of the narrowness of the column and the fine particles within it, the pump system that is incorporated into the HPLC apparatus needs to provide a high pressure to achieve satisfactory flow rates (*see* Sections 2 and 3.). The pump should be resistant to the solvents that are used in phospholipid HPLC.

### 1.2.5. The Solvent (Mobile Phase)

In isocratic separations, a single solvent or mixture of solvents is used. Alternatively, a gradient system, where the composition of the mobile phase is changed according to a special program can be used. The choice of mobile phase depends on the degree of diversity of phospholipids to be separated and on the aims of the separation.

The solvent system must be chosen carefully to suit the separation conditions and not to interfere with the detection technique. If UV detection is used, the solvents must have low absorbances in this region of the spectrum. With refractive index detection, the solvents must have a refractive index quite distinct from that of the sample components (*see* Section 1.2.7.). Solvents commonly used for phospholipid HPLC are water, aqueous buffer solutions, acids and alkalies, or organic solvents, such as acetonitrile, methanol, or chloroform, acids, and alkalies. All solvents must be of the highest spectroscopic purity and dust free *(4)*.

### *1.2.6. Sample Injection*

The ideal sample introduction method should be able to insert reproducibly and conveniently a wide range of sample volumes into the pressurized column as a sharp plug with little loss in efficiency. The injection system should possess zero dead volume to prevent loss of resolution and the sample is normally dissolved in a portion of the mobile phase to eliminate detection artifacts attributable to solvent discontinuities.

In the syringe-septum injection method, a small (10 µL) sample is introduced into the pressurized column from a high-pressure syringe through a self-sealing elastomer septum and directly onto the column packing. If the injection is directly into the separation column (i.e., no guard column), the separation column must be topped with Ballotini glass beads (30–40 µm) so as to prevent disturbances to the top of the column packing which would impair performance. The method is limited to pressures of up to $5 \times 10^6$ Pa, since at higher pressures there is sample leakage from around the syringe plunger and it becomes difficult to insert the needle into the system. This method is simple, inexpensive, and sample size can be easily varied; however, it requires a great deal of practice on the part of the operator.

For higher pressures, up to $5 \times 10^7$ Pa, a Multiport valve can be used for very precise sample loading. Such valves allow the sample to be introduced into the pressurized column without interruption to the solvent flow. The sample is first drawn into the core of the valve at atmospheric pressure; the core is then mechanically displaced and the sample pushed into the head of the column with minimal dilution. In loop injectors, a small sample loop is filled via a syringe at atmospheric pressure; then by valve switching, the sample is displaced by diverting the flow of mobile phase through the loop and onto the column.

### *1.2.7. Detection*

An ideal detector should not respond differentially according to the degree of saturation of the phospholipid molecules. It must be easy to operate and not be affected by the solvent or by changing solvent composition in the case of a gradient. Some workers recommend UV absorbance at wavelengths in the region of 195–215 nm, either directly *(5)* or after derivatization of the data to facilitate UV detection *(6)*. The main disadvantage is the differential UV absorption of phospholipids having different degrees of unsaturation. Quantitation is very difficult except with totally saturated samples or when the fatty acid composition is identical in each phospholipid class (or subclass) and a reference is used for calibration. Quantitative results cannot be obtained by integration of peaks obtained with UV detectors, because the absorbance at 200–210 nm may vary with fatty acid composition *(7,8)*. Other authors have, however, used UV peak area to make quantitative calculations *(3,9)*.

Refractive index detectors, which basically comprise a glass prism wetted by a thin layer of the column eluate, depend for their efficacy on the difference between the refractive indices of the solute and the mobile liquid phase. Saturated and unsaturated lipids do have different refractive indices, although this differential problem is less severe than with UV detection. The response of the detector is also affected by pressure fluctuations, and the use of refractometers in gradient separations is restricted to a few pairs of solvents that have virtually identical refractive indices *(1)*. A list of refractive indices of some solvents in common use is given in Table 1.

In flame ionization detection *(10)*, the solvent is evaporated and the solute converted to volatile hydrocarbons, which are combusted in a hydrogen flame. Ionization occurs and ion currents can be amplified and measured *(9)*. Large class-to-class response differences can be found.

Universal detectors such as those for phosphorus, postcolumn reaction detectors, nephelometry, infrared detectors, and mass spectroscopy are also used *(1)*. The mass spectrometer is probably the ideal detector for liquid chromatography because it is capable of providing both structural information and quantitative analysis for separated components. The mismatch between the mass flows involved in conventional HPLC, which are two or three orders of magnitude larger than can be accommodated by conventional mass spectrometer vacuum systems, requires a sampling system to be incorporated.

## 2. Materials
### 2.1. HPLC System

1. Beckman (Fullerton, CA) Model 332 Liquid Chromatographic System with a Model 110A Pump and a Model 420 Systems Controller (*see* Note 1).
2. Detector: Beckman (Fullerton, CA) 155 Variable Wavelength Detector fitted with a 20 μL, 1 cm optical path cell, set to 202 nm (*see* Notes 1, 2, and 8).

Table 1
Refractive Index of Commonly Used Solvents

| Solvent | Refractive Index |
|---|---|
| Pentane | 1.358 |
| Hexane | 1.375 |
| Cyclohexane | 1.426 |
| Carbon disulfide | 1.628 |
| 1-Chlorobutane | 1.402 |
| Diisopropylether | 1.368 |
| 2-Chloropropane | 1.378 |
| Diethylether | 1.353 |
| Chloroform | 1.443 |
| Methylene dichloride | 1.425 |
| Methylisobutylketone | 1.394 |
| Tetrahydrofuran | 1.407 |
| Acetone | 1.359 |
| 1,4-Dioxane | 1.422 |
| Ethyl acetate | 1.370 |
| 1-Pentanol | 1.410 |
| Acetonitrile | 1.344 |
| 1-Propanol | 1.380 |
| Methanol | 1.329 |
| Water | 1.333 |

Data from: Snder, L. R., in *Principles of Adsorption Chromatography,* Marcel Dekker, New York, 1968 *(11).*

## 2.2. Normal Phase Separation

### 2.2.1. Columns

A Beckman 4.6 × 250 mm column packed with 5 μm Ultrasphere Si, protected by a Guard Column of 4.6 × 45 mm packed with silica gel.

### 2.2.2. Solvents

1. HPLC-grade glass-distilled acetonitrile, chloroform, and methanol.
2. Analytical ACS-grade sulfuric acid.
3. For isocratic separation mix acetonitrile:methanol:sulfuric acid (100:3:0.5 [v/v/v]). Degas all solvents by sonication for 15 min before use.

### 2.2.3. Sample Preparation and Injection

1. Homogenizer: An all glass Dounce type homogenizer.
2. Scintered glass filter.
3. Chloroform:methanol (2:1 [v/v]).

4. 0.9% NaCl.
5. Upper phase solvent: methanol:0.9% NaCl:chloroform (48:47:7 [v/v/v]), *(5)*.
6. Chloroform:diethyl ether (1:2 [v/v]).
7. Oxygen-free nitrogen.
8. 10-μL Hamilton syringe.

### 2.3. Reverse-Phase HPLC (2,12)

Sample preparation and HPLC equipment as in Section 2.2.

#### 2.3.1. Column

Ultrasphere ODS column, 4 × 250 mm (Altex Scientific Inc., Berkeley, CA).

#### 2.3.2. Solvents

1. Solvent 1: 20 m$M$ choline chloride in methanol:water:acetonitrile (90.5:7.0:2.5 [v/v/v]). Use this solvent for the separation of phosphatidylcholine, phosphatidylethanolamine, and phosphatidylinositol species.
2. Solvent 2: 30 m$M$ choline chloride in methanol:25 m$M$/L KH$_2$PO$_4$:acetonitrile:acetic acid (90.5:7.0:2.5:0.8 [v/v/v/v]). Use this solvent for the separation of phosphatidylserine species.
3. Solvent 3: Methanol:5 m$M$ phosphate buffer, pH 7.4, ratio varies from 9:1 to 9.8:0.2 (v/v). Use this for the separation of sphingomyelin species (*see* Note 3).

### 3. Methods
### 3.1. Normal Phase
#### 3.1.1. Extraction of Lipids
#### and Sample Preparation (see also *Chapter 14*)

1. Homogenize the membrane sample in 20 vol of chloroform/methanol and filter.
2. Add 0.2 vol of 0.9% NaCl, mix and remove the upper phase, then wash the lower phase, which contains the phospholipids, three times with 0.2 vol of upper phase solvent.
3. Dry the sample under nitrogen and dissolve in chloroform/diethyl ether to 1 mg/mL.

#### 3.1.2. HPLC

1. Wash the column overnight with the solvent at a flow rate of 0.1 mL/min (*see* Note 4).
2. Apply 10 μL of sample to the column from a Hamilton syringe.

3. Adjust the flow rate of the solvent to 1.0 mL/min with a pump pressure of about $1 \times 10^7$ Pa.
4. The isocratic separation should take about 40 min (*see* Notes 5–8).

### 3.2. Reverse Phase

The sample should be prepared in chloroform/diethyl ether as in Section 3.1.1.

Run solvents 1 and 2 at about 2 mL/min and solvent 3 at 1–2 mL/min (*see* Note 9).

## 4. Notes

1. This chapter cannot provide details of the operation of specific HPLC equipment or the Detection System. This can be obtained from the appropriate manuals.
2. The detector can give information on the elution of the various classes of phospholipid, but separate chemical assay of the eluted fractions is needed for accurate quantitation *(13)*.
3. The choice of the ideal ratio from this range should be decided by experiment and depends on the source of the sphingomyelin.
4. When in continuous use, the column should be washed each night with the solvent at a flow rate of 0.1 mL/min. The entire system should be flushed out with 330 mL of methanol every month.
5. Without the use of a gradient, this isocratic system efficiently separates phospholipids, which are eluted in the following order: phosphatidylinositol, phosphatidylserine, phosphatidylethanolamine, phosphatidylcholine, and sphingomyelin.
6. There is no significant change in the fatty acid components of the phospholipids at the end of the separation.
7. Cardiolipin is not separable from the solvent front, which also contains neutral lipids, cholesterol, and cholesterol esters.
8. Disaturated phosphoglycerides, such as dipalmitoyl phosphatidylcholine, which are components, for example, of pulmonary surfactant, will not be detected on the chromatograph if they are eluted separately from the others, because the UV detection depends on the presence of double bonds.
9. Dry the fractions collected from the reverse-phase column under nitrogen and reextract. Peaks are identified by fatty acid analysis using gas-liquid chromatography (*see* Chapter 17) and quantitated by phosphorus determination (*see* Chapter 14).

# References

1. Willard, H. H., Merrit, L. L., Dean, J. A., and Sheille, F. A. (1988) High-performance liquid chromatography methods, in *Instrumental Methods of Analysis* (Willard, H. H., ed.), Wadsworth, Belmont, CA, pp. 580–613.
2. Patton, G. M., Fasulo, J. M., and Robins, S. J. (1982) Separation of phospholipids and individual molecular species of phospholipids by high performance liquid chromatography. *J. Lipid Res.* **23,** 190–196.
3. Geurts Van Kessel, W. S. M., Hax, W. M. A., Demel, R. A., and De Gier, J. (1977) High performance liquid chromatographic separation and direct ultraviolet detection of phospholipids/ *Biochim. Biophys. Acta* **486,** 524–530.
4. Wilson, K. and Goulding, K. H. (1986) Chromatographic techniques, in *A Biologist's Guide to Principles and Techniques of Practical Biochemistry,* 3rd ed., Edward Arnold, p. 235.
5. Jungalwala, F. B., Turel, R. J., Evans, J. E., and McCluer, R. H. (1975) Sensitive analysis of ethanolamine and serine containing phosphoglycerides by high performance liquid chromatography. *Biochem. J.* **145,** 517–526.
6. Hsieh, J. Y. K., Welch, D. K., and Turcotte, J. G. (1981) High pressure liquid chromatographic separation of molecular species of phosphatidic acid dimethyl esters derived from phosphatidyl choline. *Lipids* **16,** 761–763.
7. Kaduce, T. L., Norton, K. C., and Spector, A. A. (1983) A rapid isocratic method for phospholipid separation by high performance liquid chromatography. *J. Lipid Res.* **24,** 1398–1403.
8. Jungalwala, F. B., Evans, J. E., and McCluer, R. H. (1976) High performance liquid chromatography of phosphatidyl choline and sphingomyelin with detection in the region of 200 nm. *Biochem. J.* **155,** 55–60.
9. Privett, O. D. and Erdahl, W. L. (1978) An improved flame ionization detector for high performance liquid chromatography. *Anal. Biochem.* **84,** 449–461.
10. Phillips, F. C., Erdahl, W. L., and Privett, O. S. (1982) Quantitative analysis of lipid classes by liquid chromatography via a flame ionization detector. *Lipids* **17,** 992–997.
11. Snyder, L. R. (1968) *Principles of Absorption Chromatography: the separation of nonionic compounds.* Marcel Dekker, New York.
12. Jungalwala, F. B., Hayssen, V., Pasquini, J. M., and McCluer, R. H. (1979) Separation of molecular species of sphingomyelin by reversed-phase high performance liquid chromatography. *J. Lipid Res.* **20,** 579–587.
13. Bartlett, G. R. (1959) Phosphorus assay in column chromatography. *J. Biol. Chem.* **234,** 466–468.

## References

1. Willard, H. H., Merritt, L. L., Dean, J. A., and Shettle, F. A. (1982) High-performance liquid chromatography methods, in *Instrumental Methods of Analysis* (Willard, H. H., ed.), Wadsworth, Belmont, CA, pp. 580–614.

2. Patton, G. M., Fasulo, J. M., and Robins, S. J. (1982) Separation of phospholipids and individual molecular species of phospholipids by high performance liquid chromatography, *J. Lipid Res.* 23, 190–196.

3. Geurts Van Kessel, W. S. M., Hax, W. M. A., Demel, R. A., and the Gier, J. (1977) High performance liquid chromatographic separation and direct ultraviolet detection of phospholipids, *Biochim. Biophys. Acta* 486, 524–530.

4. Watson, K. and Gaultitie, K. H. (1984) Chromatographic techniques, in *A Biologist's Guide to Principles and Techniques of Practical Biochemistry*, 2nd ed., Edward Arnold, p. 84.

5. Nakagawa, Y. and Horrocks, L. A. (1983) Separation of alkenylacyl, alkylacyl, and diacyl analogues and their molecular species by high performance liquid chromatography, *J. Lipid Res.* 24, 1268–1275.

6. Hanson, V. L., Park, J. Y., Osborn, T. W., and Kiral, R. M. (1981) High-performance liquid chromatographic analysis of egg yolk phospholipids, *J. Chromatogr.* 205, 393–400.

7. Kiuchi, K., Ohta, T., and Ebine, H. (1977) High performance liquid chromatographic separation of phospholipids, *J. Chromatogr. Sci.* 15, 350–354.

8. Chen, S. S. and Kou, A. Y. (1982) Improved procedure for the separation of phospholipids by high performance liquid chromatography, *J. Chromatogr.* 227, 25–31.

9. Yandrasitz, J. R., Berry, G., and Segal, S. (1981) High performance liquid chromatography of phospholipids with ultraviolet detection, *J. Chromatogr.* 225, 319–328.

10. Sugiura, T., Nakajima, M., Sekiguchi, N., Nakagawa, Y., and Waku, K. (1983) Different fatty chain compositions of alkenylacyl, alkylacyl, and diacyl phospholipids in rabbit alveolar macrophages, *Lipids* 18, 125.

# CHAPTER 16

# Measurement
# of Cholesterol in Membranes

## Hafez A. Ahmed

## 1. Introduction

Cholesterol is an amphipathic molecule of great importance in biology. It can be measured by enzymic and chemical methods, directly or indirectly following extraction from tissues or membranes. Before determination, cholesterol esters must be hydrolyzed using either chemical methods or cholesterol ester hydrolase (EC 3.11.13) (*see* Note 1). Enzymic methods for cholesterol determination are more sensitive and convenient and are generally used in preference to chemical methods.

Cholesterol oxidase (EC 1.1.3.6) catalyzes the stoichiometric conversion of cholesterol into cholest-4-en-3-one with the liberation of hydrogen peroxide. This can be coupled to a color-producing system and quantified. Other enzymic methods measure oxygen consumed by an oxygen-sensitive electrode. Alternatively, cholest-4-en-3-one can be quantitatively extracted using 2,2,4-trimethylpentane (isooctane) and measured spectrophotometrically *(1)*.

Chemical methods for cholesterol assay depend on the reaction of cholesterol with strong concentrated acids to form colored cholestapolyenes and carbonium ions (Lieberman-Burchard reaction). Acetic acid and acetic anhydride are used as solvents and dehydrating agents whereas sulfuric acid is used as an oxidizing and dehydrating agent. The reaction is enhanced by addition of various metal ions *(2)*. Francey et al. *(3)* used ethyl acetate as a solvent for ferric chloride and completely extracted cholesterol in ethanol.

From: *Methods in Molecular Biology, Vol. 19: Biomembrane Protocols: I. Isolation and Analysis*
Edited by: J. M. Graham and J. A. Higgins Copyright ©1993 Humana Press Inc., Totowa, NJ

## 2. Materials

### 2.1. Enzymic Method

1. Phosphate buffer: 8.0 g $Na_2HPO_4$ (anhydrous) and 4.0 g $KH_2PO_4$ (anhydrous) dissolved in 1 L of water and adjusted to pH 6.8.
2. Ethanolic potassium hydroxide: Mix 4 mL of $12M$ potassium hydroxide with 96 mL of ethanol.
3. $KH_2PO_4$: 120 g/L.
4. Wetting agent: Lutensol 2% (w/v) (*see* Note 2).
5. Cholesterol oxidase: 4 U/mL.
6. Oxidase reagent: Mix 425 mL of phosphate buffer (pH 6.8) with 50 mL of $KH_2PO_4$ and 25 mL of wetting agent. Add 30 U of cholesterol oxidase.
7. Isooctane (2,2,4, trimethylpentane) (*see* Note 3).
8. Cholesterol standard solution: 6.25 m$M$ in ethanol.
9. 12-mL Screw-cap glass tubes.
10. Bench centrifuge.
11. UV spectrophotometer.

### 2.2. Chemical Method (3)

1. Ethanol: absolute or 95%.
2. Ferric chloride in ethyl acetate: Dissolve 100 mg of ferric chloride hexahydrate in 100 mL of ethyl acetate, and store the solution in an amber glass bottle (stable for at least 6 mo).
3. Concentrated sulfuric acid.
4. Cholesterol standard solution 25 mmol/L in ethanol.
5. 12-mL Screw-capped tubes.
6. 15-mL Quickfit glass-stoppered tubes.
7. Bench centrifuge.
8. Visible wavelength spectrophotometer.

## 3. Methods

### 3.1. Enzymic Method

1. Add 2.0 mL of oxidase reagent to 20 μL of membrane sample (*see* Note 4) or cholesterol standard in ethanol in a 12-mL screw-capped glass tube.
2. Incubate the tubes for 15 min at 50°C.
3. Stop the reaction by addition of 2.0 mL of ethanolic potassium hydroxide, followed by 5.0 mL of isooctane.
4. Cap the tube tightly, shake vigorously for 1 min, and centrifuge at 1000$g$ for 10 min.
5. Measure the absorbance of the isooctane (upper layer) at 232 nm using a quartz 1-cm cell against a blank treated in exactly the same way but lacking cholesterol oxidase.

6. Prepare a standard curve using between 25 and 250 nmol cholesterol. The expected absorbances at 232 nm are between 0.12 and 0.87, respectively.

## 3.2. Chemical Method

1. Add 1.9 mL of ethanol to 0.1 mL of membrane preparation in a screw-capped tube.
2. Shake vigorously for 15 s and centrifuge at 1000g for 10 min (*see* Note 5).
3. Place 0.5 mL of the ethanol supernatant into a 15-mL Quickfit glass-stoppered tube.
4. Add 2.0 mL of ferric chloride solution, mix, and stand for 10 min.
5. Cool the tubes by immersion in ice and add 2.0 mL of concentrated sulfuric acid very slowly over a period of about 5 s while continually swirling the tube.
6. Allow the tube to stand for 10 min and read the absorbance at 560 nm against the reagent blank. Plot a standard curve using cholesterol solutions of up to 25 m$M$. An absorbance of 0.3 is expected for a cholesterol solution of 5.17 m$M$.

## 4. Notes

1. Cholesterol ester hydrolase is ineffective against some cholesterol esters and chemical hydrolysis must be used.
2. The use of an efficient wetting agent is essential in enzymic methods for cholesterol assay, especially of membranes. The wetting agent should not have a high absorbance at 232 nm. Triton-X-100 is an efficient wetting agent but has a high absorbance at 232 nm when extracted into isooctane. Lutensol has a minimal absorbance and is suitable for these methods.
3. Isooctane is used as a solvent for cholest-4-en-3-one because it has a low molar absorption at 232 nm. Other solvents such as petroleum spirit may have high absorbance at this wavelength. This should be checked if alternative solvents are used.
4. Cholesterol esters must be hydrolyzed before assay by the enzymic method. Boil the samples in 200 μL of ethanolic potassium hydroxide. Cool the samples and add aliquots to cholesterol oxidase and assay as for free cholesterol. Although the buffering capacity of the assay mixture is high, the pH of the oxidase reagent should be checked after addition of ethanolic potassium hydroxide and readjusted to 6.8 if necessary.
5. Cholesterol in lipid extracts or after separation by thin layer chromatography (*see* Chapters 14 and 17) may be assayed either by the enzymic or chemical methods. The cholesterol-containing samples should be dissolved in a small amount of ethanol (10–100 μL) before proceeding as described earlier.

# References

1. Trinder, P. (1981) Oxidase determination of plasma cholesterol as cholest-4-en-3-one using iso-octane extraction. *Ann. Clin. Biochem.* **18,** 64–70.
2. Stein, E. A. (1987) Lipids, lipoproteins and apolipoproteins, in *Fundamentals of Clinical Chemistry* (Tietz, N. W., ed.), Saunders, Philadelphia, PA, pp. 448–481.
3. Franey, R. J. and Amador, E. (1968) Serum cholesterol measurement based on ethanol extraction and ferric chloride sulphuric acid. *Clin. Chim. Acta* **21,** 255–263.

CHAPTER 17

# Analysis of Fatty Acids
# by Gas Liquid Chromatography

## Ian J. Cartwright

## 1. Introduction

The fatty acids (FA) of animal, plant, and microbial origin are predominantly unbranched aliphatic chains with an even number of carbon atoms and a single carboxyl group. FA can be classified as saturated, monounsaturated, or polyunsaturated (Fig. 1). Saturates have all single carbon-to-carbon bonds. Monounsaturates contain a single *cis* double bond, whereas polyunsaturates have two or more. A list of the commonly occurring natural FA in animal tissues is shown in Table 1. In general, FA do not exist as free carboxylic acids in living cells. In animals, they are mainly esterified into membrane phospholipids (phosphoglycerides and sphingolipids) and cytosolic neutral lipids (triglycerides and cholesteryl esters). In order to analyze the FA components of cell lipids, they must first be converted into nonpolar, volatile derivatives before they can be separated and quantified by the rapid and highly sensitive technique of gas liquid chromatography (GLC).

GLC can be regarded as a form of partition chromatography, in which the stationary phase is a liquid and the mobile phase a gas (1). A sample is injected into the gas phase, where it vaporizes and moves into the liquid phase, which is held on an inert support material within a column. Individual sample components spend different times in the gas and liquid phases, depending on their relative affinities for the latter. Each separate component emerges from the end of the column,

From: *Methods in Molecular Biology, Vol. 19: Biomembrane Protocols: I. Isolation and Analysis*
Edited by: J. M. Graham and J. A. Higgins Copyright ©1993 Humana Press Inc., Totowa, NJ

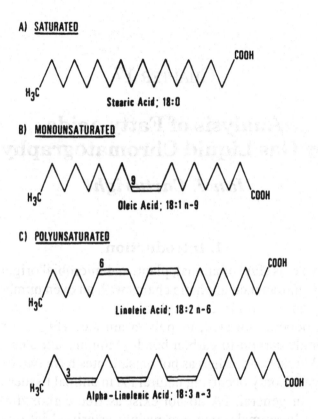

Fig. 1. Structural classification of fatty acids.

and is detected by ionization in a diffusion flame of hydrogen and air. This generates an electrical signal that is proportional to the component concentration, and this can be amplified and directed to a continuous chart recorder. Peak areas can be measured from the recorder in order to allow quantitation of components.

The aim of this chapter is to describe a simple GLC method for the separation and quantitation of the FA components of individual membrane and cytosolic lipids from animal cells.

## 2. Materials
### 2.1. Glassware

1. Pyrex glass tubes (6-mL capacity), with Teflon-lined screw caps.
2. Disposable glass Pasteur pipets.

Table 1
Nomenclature of Some Common Natural Fatty Acids

| Common name | Synonym[a] | Abbreviation[b] |
|---|---|---|
| Myristic | Tetradecanoic | 14:0 |
| Palmitic | Hexadecanoic | 16:0 |
| Stearic | Octadecanoic | 18:0 |
| Arachidic | Eicosanoic | 20:0 |
| Behenic | Docosanoic | 22:0 |
| Lignoceric | Tetracosanoic | 24:0 |
| Cerotic | Hexacosanoic | 26:0 |
| Palmitoleic | 7-Hexadecenoic | 16:1 n-9 |
| Oleic | 9-Octadecenoic | 18:1 n-9 |
| Gadoleic | 11-Eicosenoic | 20:1 n-9 |
| Erucic | 13-Docosenoic | 22:1 n-9 |
| Nervonic | 15-Tetracosenoic | 24:1 n-9 |
| Linoleic | 9,12-Octadecadienoic | 18:2 n-6 |
| Alpha-linolenic | 9,12,15-Octadecatrienoic | 18:3 n-3 |
| Gamma-linolenic | 6,9,12-Octadecatrienoic | 18:3 n-6 |
| Arachidonic | 5,8,11,14-Eicosatetraenoic | 20:4 n-6 |
| Adrenic | 7,10,13,16-Docosatetraenoic | 22:4 n-6 |
| Timnodonic | 5,8,11,14,17-Eicosapentaenoic | 20:5 n-3 |
| Clupanodonic | 4,7,10,13,16,19-Docosahexaenoic | 22:6 n-3 |

[a]Double bonds numbered from the carboxyl end of the molecule.

[b]Abbreviated fatty acid formulae showing the number of carbon atoms (e.g., 12,18,22), the number of double bonds (e.g., 0,1,4), and the position of the first double bond in relation to the n-carbon or methyl end of the molecule (e.g., n-3,n-6).

## 2.2. Apparatus

1. Long-tipped, calibrated, positive-displacement pipets (0.1-, 0.2-, 1.0-, and 5.0-mL capacities) for accurate pipeting of solutions of lipids in organic solvents.
2. A thermostatically controlled, aluminum heating block (to accommodate the aforementioned 6-mL Pyrex glass tubes) for incubation of lipid samples (in organic solvents) during drying and FA methylation procedures.
3. A microsyringe (10-μL capacity), for injection of fatty acid methyl ester (FAME) solutions into the column of the gas chromatograph (GC).
4. A 3-m silanized glass GLC column with a 2-mm internal diameter.
5. A GC equipped with a hydrogen flame ionization detector (FID), continuous chart recorder, and digital integrator. A wide variety of GC models are commercially available, and the major manufacturers and suppliers are: Chrompack (Millharbour, UK), Fisons Instruments

(Crawley, UK), Hewlett-Packard (Wokingham, UK), Perkin-Elmer (Beaconsfield, UK), Philips Scientific (Cambridge, UK), and Siemens AG (Karlsruhe, Germany). GLC glass columns and packing material are also available from these companies.

6. A supply of nitrogen (oxygen-free) for evaporation of organic solvents from lipid samples.
7. Supplies of compressed air and hydrogen for the FID and high purity argon (the carrier gas). Each supply must be equipped with a moisture trap.

## 2.3. Reagents

1. Boron trifluoride (12–14% [w/v] in methanol). After each use, flush out the air in the reagent bottle with nitrogen, reseal, and store at 4°C, in the dark. Under these conditions, the reagent is stable for about 3 mo. The use of old or too concentrated solutions may result in the production of artifacts and significant losses of polyunsaturated FA (*see* Note 1).
2. Chloroform (analytical grade).
3. Dichloromethane (analytical grade).
4. Ethyl acetate (analytical grade).
5. Hexane (analytical grade), containing the antioxidant, butylated hydroxytoluene (BHT) (150 µg/mL).
6. Methanol (analytical grade).
7. 1$M$ NaOH (aq.).
8. GLC column packing material: 10% Silar 10C (stationary phase) supported on Gaschrom Q (100–120 mesh).
9. Pyrex glass wool (acid-washed and silanized).

## 2.4. Standards

1. A range of standard FA commonly encountered in animal tissues: 16:0, 18:0, 20:0, 22:0, 24:0, 26:0, 16:1 n-9, 18:1 n-9, 24:1 n-9, 18:2 n-6, 20:3 n-6, 20:4 n-6, 22:4 n-6, 20:5 n-3, 22:5 n-3, and 22:6 n-3 (*see* Table 1 for common names).
2. The internal standard FAME solution: methyl heneicosanoate (21:0) (0.15 m$M$, in ethyl acetate). Store this at –20°C.

## 3. Methods

### 3.1. Methylation of Fatty Acids

Before FA can be analyzed by GLC, they must be converted to a more volatile form. The derivatives most commonly prepared are the FA methyl esters (FAME). It is not necessary to hydrolyze acyl lipids to obtain the free FA, before preparing the methyl esters, since most of these can be transesterified directly. The acid-catalyzed reaction employing boron trifluoride-methanol ($BF_3$) can be used to transesterify most of the acyl lipid classes in the membrane and cytosolic components of animal cells *(2)*. Using this method, it is possible to transesterify acyl lipids on TLC adsorbents, without the need for prior elution. This clearly simplifies the methodology and reduces the possibility of contamination. By varying the reaction conditions, quantitative recovery of FAME can be achieved from phospholipids: diphosphatidylglycerol (DPG), phosphatidylcholine (PC), phosphatidylethanolamine (PE), phosphatidylinositol (PI), phosphatidylserine (PS), sphingomyelin (SM); from neutral lipids: triglycerides (TG) and cholesteryl esters (CE); and from glycolipids: ceramide monohexosides (CM) and gangliosides (GS).

1. For total lipid extracts of membranes or whole tissue, transfer the sample (in chloroform solution) to a glass tube (with screw-cap) and incubate at 30°C (in the heating block); evaporate to dryness under a stream of nitrogen and add 1.5 mL of $BF_3$ reagent. For samples of individual acyl lipid classes separated by TLC (*see* Chapter 14), scrape the band of TLC adsorbent containing the lipid fraction into a glass tube (with screw-cap). Then add a sufficient volume of $BF_3$ reagent to allow complete drenching of the silica gel in each tube.
2. Add 500 µL of dichloromethane to the total lipid, TG, and CE tubes only, and vortex-mix. This improves the solubility of the nonpolar acyl lipid in the polar reagent solution and greatly reduces transesterification times.
3. Flush out the remaining air in all reaction tubes with nitrogen, and replace and tighten the screw-caps. Shake the tubes vigorously for 1 min.
4. Incubate (in the heating block) the reaction tubes containing free FA, DPG, PC, PE, PI, PS, TG, and CE at 65°C for 1 h. Remove these tubes and heat the block to 90°C. Then incubate total lipid for 1 h and SM, CM, and GS for 3 h (*see* Note 2).
5. Allow all reaction tubes to cool for 5 min (*see* Note 3), then carefully unscrew the caps and add 3 mL of hexane (containing antioxidant), followed by 1.5 mL of $1M$ NaOH (*see* Note 4).
6. Replace the screw-caps and shake the tubes vigorously for 2 min, and allow the aqueous and organic phases to separate. If silica gel emulsions form in the organic phase, these must be separated by centrifugation at 1000*g* for 10 min.
7. Using disposable glass Pasteur pipets, carefully transfer the upper organic phase (containing FAME) in each tube to separate, preweighed, clean, glass tubes (with screw-caps). Avoid transfer of any of the aqueous layer.

8. Re-extract the aqueous layer with 3 mL of hexane, and pool the organic phases for each sample.
9. Incubate the tubes containing the hexane extracts at 30°C, and evaporate the FAME to dryness under a stream of nitrogen. Reweigh the tubes and determine the dry weight of FAME in each sample.
10. Redissolve the dry FAME samples (1–5 mg) in 200 µL of chloroform, flush the remaining air from the tubes with nitrogen, and replace and tighten the screw-caps (*see* Note 5).
11. Store the FAME solutions at –70°C prior to GLC analysis. Under these anoxic conditions there are no significant losses of polyunsaturated FA as a result of autooxidation for up to 4 wk. However, if quantitation is required, FAME should ideally be prepared as close to the analysis time as possible.

## 3.2. GLC of Fatty Acids

A detailed account of the theory of GLC and of the design and operation of GCs will not be attempted here. Only the most basic procedures and methods involved in GLC will be presented, and the reader should refer to the more specialized texts for further information *(1,3)*.

### 3.2.1. Packing of Columns

One of the most important components of a GC is the column and its packing material, which allows the separation of FAME. Although wall-coated, capillary columns that give very high resolution are now available, these have the disadvantage of the difficulties in identification of large numbers of minor FA components, which may have little biochemical relevance. Analysis with packed columns, on the other hand, gives simplified chromatograms, but with much of the required information. Therefore, the methods described for the quantitation of FA in this chapter will be confined to the use of packed columns. Although prepacked columns are available commercially, it is less expensive to pack empty columns in the laboratory, using the following simple technique.

1. Attach a piece of Tygon tubing to the exit end of the column and apply a vacuum.
2. Insert a small plug (about 1 cm) of Pyrex glass wool (acid-washed and silanized) at the inlet end, and gently tap the column until the plug lodges at the exit end. This plug prevents the entrance of packing material into the FID (*see* Note 6).

3. Attach a funnel to the exit end and add small amounts of packing material (10% Silar 10C on Gaschrom Q), while gently tapping the column. Pack the column up to 5 cm from the inlet end, and insert another glass wool plug to consolidate it.
4. Put the packed column into the GLC oven and attach at the inlet end only. Turn on the carrier gas (argon) supply and adjust the flow rate to 45 mL/min (*see* Note 7).
5. Condition the packing material by heating the column at 250°C for 24 h. After conditioning, the column is ready for FAME separation, and should have an operating life of over 1 yr, if used carefully (*see* Notes 8–10).

### 3.2.2. GLC Operating Conditions

The following GLC operating conditions allow the resolution of FA with 16–26 carbon atoms, including those containing up to 6 double bonds. A preprogrammed, thermal gradient is applied during each analysis, and as the oven temperature increases, the FA emerge from the column generally in order of increasing chain length and number of double bonds (*see* Fig. 2).

1. Adjust the carrier gas (argon) flow rate to 45 mL/min.
2. Set the FID temperature to 250°C.
3. Set the initial oven temperature to 195°C and hold this for 8 min.
4. Set the temperature gradient to 7°C/min.
5. Set the final oven temperature to 245°C and hold this for 20 min.

### 3.2.3. Sample Injection

Good sample injection technique is important, both for quantitative recovery and optimal resolution of FAME.

1. Prior to analysis, check the septum for leaks and replace if necessary. Septum bleed may lead to appreciable loss of sample upon injection.
2. Wash out the microsyringe barrel with at least three changes of ethyl acetate.
3. Take up 5 μL of sample (in ethyl acetate) in the microsyringe. Ensure that no air bubbles are aspirated with the sample solutions, as these will cause gradual oxidative degradation of the liquid phase if repeatedly injected into the column (*see* Note 11).
4. Insert the microsyringe needle rapidly but steadily through the septum and into the column, until it reaches a predetermined depth (about 5 mm) below the surface of the packing material.
5. Press the plunger firmly in, leave the needle in place for about 2 s, and then rapidly remove the syringe.

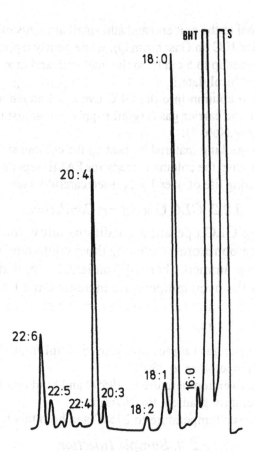

Fig. 2. GLC separation of the FAME derivatives of human erythrocyte PS on a glass column (3 m × 2 mm id) packed with 10 % Silar 10C (stationary phase) on Gaschrom Q (100–120 mesh) support. The flow rate of the carrier gas (argon) was 45 mL/min. The oven of the Pye Unicam Model 104 GC was maintained initially at 195°C for 8 min, and was then temperature-programmed to 245°C at 7°C/min. The final temperature was held for 15 min. The two peaks eluting prior to 16:0 were the antioxidant (BHT) and the carrier solvent (S).

6. Wash out the microsyringe barrel with at least three changes of ethyl acetate.

### 3.2.4. Peak Identification

Fatty acids in an unknown sample are provisionally identified by direct comparison of the relative retention times (RRT) of their methyl esters with those of FA standards separated on the same column and under identical GLC operating conditions. The RRT of a FAME is the amount of time it has spent on the column relative to that of a chosen standard, commonly occurring component (usually 18:0). RRTs are determined as follows.

1. Prepare standard mixtures of saturated FA (16:0 18:0, 20:0, 22:0, 24:0, 26:0), monounsaturated FA (16:1 n-9, 18:1 n-9, 24:1 n-9), and polyunsaturated FA (18:2 n-6, 20:3 n-6, 20:4 n-6, 20:5 n-3, 22:4 n-6, 22:5 n-3, 22:6 n-3). Use 500 μg of each FA and dissolve each mixture in 1 mL of methanol. Also add 500 μg of 18:0 to the mono- and polyunsaturated FA mixtures.
2. Add 1.5 mL of $BF_3$ reagent to each standard mixture, and prepare methyl esters of the free FA as described in Section 3.1.
3. Dry the FAME under nitrogen and redissolve each standard mixture in 500 μL of ethyl acetate.
4. Inject 5 μL of each standard mixture into the column, and separate the FAME under the standard GLC operating conditions (*see* Section 3.2.2.). If necessary, redilute the mixtures in ethyl acetate and reinject until all FAME peaks are on scale on the chromatogram.
5. From the chromatogram, measure the retention time of each FAME on the column, i.e., the time from the point of emergence of the carrier solvent peak (ethyl acetate) to the maximum peak height of each FAME.
6. Divide the retention time of each FAME by that of the chosen standard FA (18:0). Use these RRTs to identify FA in unknown samples separated under identical GLC operating conditions.

### 3.2.5. Calibration of FID Response

The areas under the peaks on a GLC trace are within limits linearly proportional to the amount (by weight) of the FAME components eluting from the column. Although modern FIDs usually have sufficient linear range to cope with most analytical problems, quantitative errors may occur if columns are heavily overloaded with sample. It is therefore advisable to calibrate the FID response with standard FAME mixtures, as described in the following (*see* Note 12).

1. Prepare standard mixtures of saturated, monounsaturated, and polyunsaturated FAME, as described in the Peak Identification section (*see* Section 3.2.4.). Dissolve each FAME mixture to a known concentration (1–10% [w/v]) in ethyl acetate.
2. Inject suitable aliquots of the standard solutions into the GLC column, such that the FAME peak heights range from 10–100% of full scale deflection on the chart recorder, for each attenuation setting, under the standard GLC operating conditions (*see* Section 3.2.2.).

3. Using the electronic integrator attached to the GLC, measure the peak area (A) of each FAME component in each run, and divide the values by the corresponding amount of sample injected (in μg) to calculate the response factor, $r = A/\mu g$ sample, for each component. The range in which $r$ is constant gives the range of linear response of the FID. Therefore, when quantitative analysis of an unknown sample is required, adjust the sample load so that it falls within the linear response range.

### 3.2.6. Quantitation of FA

It is possible to determine the absolute amounts of FA in a sample by the addition of a known amount of an internal standard FAME, prior to GLC analysis. The acyl lipid from which the FA were derived can also be quantified by this method. The internal standard FA must be selected to meet three important criteria. First, it should not normally be present in tissue lipids. Second, it must elute from the GLC column, adequately separated from all the sample FA components. Third, it should be stable and relatively nonvolatile, to allow for storage of standard solutions for extended periods of time. Heneicosanoic acid (21:0) methyl ester fits the above criteria, and can be used to quantify FA and acyl lipids as described below.

1. Prepare 500 μg of 18:0 methyl ester using $BF_3$ reagent, and evaporate to dryness under nitrogen.
2. Redissolve the dry 18:0 methyl esters in 500 μL of internal standard solution (0.15 m$M$ 21:0 methyl ester in ethyl acetate).
3. Set the attenuation on the GLC to 500×, inject 5 μL of the 18:0/21:0 methyl ester solution into the column, and separate the FAME under the standard GLC operating conditions (*see* Section 3.2.2.). If necessary, reinject the sample and adjust the attenuation to give a peak deflection for 21:0 of approx 75% of full scale. Maintain this attenuation setting for all subsequent FAME sample analyses. It may also be necessary to redilute the 18:0/21:0 sample in the internal standard solution and reinject, until the 18:0 peak is on scale.
4. Determine the retention time of 21:0 relative to 18:0.
5. For GLC analysis of FA from samples of total acyl lipid and individual acyl lipid classes (DPG, PC, PE, PI, PS, SM, CE, free FA, TG, CM, and GS), dry the chloroform solutions of FAME under nitrogen and redissolve (1–2% [w/v]) in a known volume of internal standard solution.
6. Inject 5 μL of sample onto the column, and analyze under standard GLC operating conditions (*see* Section 3.2.2.). Each sample analysis should take about 40 min. During each analysis, the remaining sample solution

should be maintained on ice to minimize evaporation of the carrier solvent.

7. If necessary, redilute the sample in a known volume of internal standard solution, and reinject until all sample FAME peaks are on scale.

8. Identify the sample FAME peaks on the GLC trace, by comparing their retention times (relative to 18:0) with those of the standard FAME, determined in Section 3.2.4. Also identify the internal standard FAME peak (*see* Note 13).

9. Determine the peak areas of the sample FAME and the internal standard FAME, using the digital integrator (*see* Note 14).

10. Sum the peak areas of all the sample FAME, and calculate the relative percentage FA composition, using Eq. 1, where $A_X$ is the area of a given peak (X), and $A_1$ to $A_N$, the sum of all the peak areas.

$$\text{Area } \% \ X = (A_X/A_1 + A_2 + \ldots \ldots A_N) \times 100 \tag{1}$$

11. Determine the absolute amount (molar and weight basis) of each FAME component in 5 μL of sample, by relating its peak area to that of the internal standard (21:0), the absolute amount of which is known. Then calculate the absolute amounts of FAME in the total volume of sample solution.

12. Determine the absolute amount of acyl lipid in the whole sample, by dividing the total number of moles of FA by the number of FA esterified per mole of lipid class (i.e., 1 FA for CE, SM, CM, GS; 2 FA for PC, PE, PI, PS; 3 FA for TG; and 4 FA for DPG).

13. Calculate the concentrations of FA and acyl lipid per unit weight of membrane or whole tissue from which each sample was derived. Alternatively, if protein assays of membrane suspensions or whole tissue homogenates have been performed, express the results as per unit weight of total protein (*see* Chapter 18).

## 4. Notes

1. Because of the toxicity of methanol and the irritant nature of $BF_3$, it is advisable to pipet the esterification reagent in a fume cupboard.

2. Transesterification of the FA in SM and glycolipids requires a higher temperature and longer reaction time, since the amide linkages of SM and glycolipid (CM and GS) are more resistant to derivatization than the ester linkages of the phosphoglycerides and neutral lipids. Clearly the reaction time and temperature must also be increased for total lipid, if adequate transesterification of its SM and glycolipid components are required.

3. Do not open hot reaction tubes immediately after transesterification, since considerable pressure may build up (especially at 90°C) and this

could lead to significant loss of sample. Allow sufficient cooling time (5 min).

4. Because of the highly flammable nature of hexane, the extraction and drying of FAME should be performed in a fume cupboard.

5. It is important to store all FAME solutions under a blanket of nitrogen (oxygen-free), at $-70°C$, prior to GLC analysis, since polyunsaturated FA rapidly autooxidize in the presence of air. Autooxidative damage to FAME during sample preparation and storage is minimized by the presence of the antioxidant (BHT) in the samples. This compound does not interfere with GLC analysis, since it is relatively volatile and tends to elute immediately after the carrier solvent peak, ahead of the FAME (*see* Fig. 2).

6. Occasionally, when plugging the end of a column, it may be necessary to suck through a small volume of methanol to help carry the glass wool to the exit. After plugging the exit, put the column into the GLC oven and connect it at the inlet end only. Then turn on the carrier gas supply, adjust the flow rate to 45 mL/min, and heat the column at 200°C for 15 min, in order to drive off the methanol.

7. The flow of carrier gas should never be allowed to stop, as long as the column is being heated, or damage to the liquid phase will result.

8. If columns are packed too tightly, the injection syringes may block easily; if they are too loosely packed, the separation of FAME is poor.

9. Occasionally, after column conditioning, the packing material may settle a few centimeters. In this case, more packing material should be added as required, and the column reconditioned for 2–3 h.

10. In regular use, the stationary phase may deteriorate, owing to introduction of trace amounts of polar impurities with the samples. These may slowly react with the liquid phase and thus alter the separation characteristics of the column. Fortunately, most damage is usually confined to the first few centimeters from the inlet end. This can be replaced periodically by fresh packing material, and column life extended considerably.

11. The FAME carrier solvent (ethyl acetate) is highly flammable, and should be used in a well ventilated area, away from naked flames and sources of electrical discharge.

12. When GLC is routinely used for FA analyses, it is important to establish a quality control system, in order to check for instrumentation faults and deterioration of reagents and column packing material, which may introduce errors into the final quantitative result *(4)*. The GLC procedures should be checked at regular intervals by running primary standard FAME mixtures of accurately known composition through the column. FAME mixtures made originally to the specifications of the

National Institutes of Health (USA) are available from a number of commercial sources. The results of quality control checks should be evaluated statistically, and if necessary, appropriate measures taken to improve the accuracy and precision of the final analytical result.

13. Under the standard GLC operating conditions (*see* Section 3.2.2.) the internal standard FAME (21:0) should run approximately midway between 18:2 n-6 and 22:0, on the chromatogram.

14. Although digital integration is certainly the most accurate method of peak area measurement, some manual methods can give acceptable results, providing the peaks are symmetrically shaped. A simple method is to measure the peak height and the peak width at half the peak height; and then multiply the two values, to give the peak area.

## References

1. Christie, W. W. (1973) Chromatographic and spectroscopic analysis of lipids: General principles, in *Lipid Analysis*, Pergamon, Oxford, pp. 42–84.

2. Morrison, W. R. and Smith, L. M. (1964) Preparation of fatty acid methyl esters and dimethyl acetals from lipids with boron-fluoride methanol. *J. Lipid Res.* **5**, 600–608.

3. Kates, M. (1986) Separation of lipid mixtures, in *Isolation, Analysis and Identification of Lipids* (Burdon, R. H. and van Knippenberg, P. M., eds.), Elsevier, Amsterdam, pp. 254–278.

4. Naito, H. K. and David, J. A. (1984) Laboratory considerations: Determination of cholesterol, triglyceride, phospholipid and other lipids in blood and tissues, in *Lipid Research Methodology* (Story, J. A., ed.), Liss, New York, pp. 1–76.

# CHAPTER 18

# Chemical Assays for Proteins

## David J. Winterbourne

## 1. Introduction

The choice of a protein assay for membranes and membrane proteins is dependent on a number of considerations: sensitivity, specificity (both with respect to variation between proteins and interference by nonprotein components), and simplicity.

The most common problem encountered in membrane studies is the interference produced by sample components such as detergents, lipids, high concentrations of dissociating agents, or density gradient materials. This interference, to which the widely used Lowry assay is particularly sensitive, may be overcome by precipitating the proteins (1), by extracting the interfering substances with appropriate solvents (2), or by sedimenting the membranes after dilution of the gradient material. The protein then requires solubilization before analysis. Even when interfering components are not present, solubilization of membrane proteins is a requirement for most solution assays.

The problem of variable response between proteins is characteristic of all assays that are dependent, at least in part, on amino acid side chain reactivity, and where measurements are made relative to a different standard protein (most of the common laboratory assays). The Lowry (3) and related bicinchoninic (4) assay are less susceptible to this source of error than the Bradford (5) Coomassie blue assay. Nonetheless, large errors can occur with purified membrane proteins determined by the Lowry assay using bovine serum albumin as a standard. In these cases it may be necessary to use an absolute protein assay such as quantitative amino acid analysis (6).

From: *Methods in Molecular Biology, Vol. 19: Biomembrane Protocols: I. Isolation and Analysis*
Edited by: J. M. Graham and J. A. Higgins Copyright ©1993 Humana Press Inc., Totowa, NJ

Two assays are described here, neither of which require membrane solubilization: First, a method for the Lowry assay, modified to permit simple determination of membrane proteins *(7)*, is described. The advantages of the Lowry assay are its widespread use, allowing comparison between laboratories, and the fairly low variation in response between different proteins. The second assay depends on the binding of Coomassie blue R250 to precipitated proteins retained on filter paper. This assay is extremely resistant to interference by other chemicals, most of which are extracted during the staining procedure while the protein is fixed to the paper support. Interprotein variability, particularly for bovine serum albumin and colored proteins, is less than with the Bradford assay, which relies on spectral changes of Coomassie blue G250 bound to proteins in solution.

## 2. Materials

### 2.1. Modified Lowry Assay (7)

1. Alkaline sodium dodecyl sulfate solution: 2% (w/v) $Na_2CO_3$, 1% (w/v) sodium dodecyl sulfate, 0.4% (w/v) NaOH, and 0.16% (w/v) sodium tartrate. Stable indefinitely at room temperature.
2. Cupric solution: 4% (w/v) $CuSO_4 \cdot 5H_2O$. Stable indefinitely at room temperature.
3. Reagent A: 100 parts alkaline sodium dodecyl sulfate solution to 1 part cupric solution. Prepare on the day of the assay.
4. Reagent B: Folin-Ciocalteu phenol reagent, dilute with an equal volume of water.
5. Standard protein solution: bovine serum albumin at 100 µg/mL
6. Test tubes (e.g., 12 × 75 mm plastic disposable tubes).
7. Spectrophotometer capable of measuring absorbance at 660 nm and 750 nm.

### 2.2. Coomassie Blue Staining Assay

1. Stock Coomassie blue solution: 0.4 g Coomassie brilliant blue R250 dissolved in 250 mL ethanol and 630 mL distilled water. This solution is stable indefinitely at room temperature.
2. Working Coomassie blue solution: 12 mL glacial acetic acid added to 88 mL of the stock solution. Final composition of the working dye solution is 0.04% (w/v) Coomassie blue, 25% (v/v) ethanol, and 12% (v/v) acetic acid. The solution should be filtered after preparation and is stable at room temperature for at least 1 mo. With time, some Coomassie blue may precipitate, however this has little effect on the assay, as standards are always carried on the same filter sheet.

3. Whatman 3MM filter paper marked in pencil with a grid of 1 cm$^2$ squares. The grid is conveniently prepared by drawing lines using a plastic ruler notched at 1 cm intervals. Larger squares will accommodate larger sample volumes but with increased background absorbance.
4. Standard protein solution: bovine serum albumin or bovine γ-globulin at 1 mg/mL.
5. 10% (w/v) Trichloracetic acid (TCA) solution.
6. Destaining solution: 10% (v/v) ethanol, 5% (v/v) glacial acetic acid.
7. Test tubes: e.g., 12 × 75 mm plastic disposable tubes.
8. Eluant: 1$M$ potassium acetate in 70% (v/v) ethanol.
9. Spectrophotometer capable of measuring absorbance at 590 nm.

# 3. Methods

## 3.1. Modified Lowry Assay (see Note 1)

1. Mix 1 mL of reagent A with 0.3 mL aliquots of samples and standards containing 3 to 30 μg protein.
2. After 10 min at room temperature add 0.1 mL of reagent B, mixing vigorously immediately after or during addition (*see* Note 2).
3. The absorbance at 660 nm should be measured after 45 min and within 2 h of step 2 (*see* Note 3).

## 3.2. Coomassie Blue Staining Assay

1. Spot 1- to 8-μL aliquots of samples and standards into the center of squares on the sheet of Whatman 3MM. The range of protein masses should be between 0.5 and 8 μg. Blank squares should be left for determination of background staining (*see* Notes 4 and 5).
2. Fix, if necessary (*see* Note 6), by immersion in 10% TCA for 15 min.
3. Transfer the sheet to a bath of working dye solution, and stain for 1 h (*see* Note 7).
4. Destain with three changes of 10% ethanol, 5% acetic acid for 10 min each.
5. Dry the sheet at room temperature or in an oven at 80°C.
6. Cut the grid into its component squares and place each square, cut in half, in a test tube. Add 1 mL of 1$M$ potassium acetate in 70% ethanol to elute the bound dye.
7. After 1 h, measure the absorbance at 590 nm. Protein concentrations in the samples can be determined by interpolation of the linear relationship between absorbance and standard protein mass (Fig. 1) (*see* Notes 8 and 9).

# 4. Notes

1. This modified assay is resistant to interference by up to 0.2$M$ sucrose and 2.5 m$M$ EDTA. Higher concentrations do interfere, as do a large num-

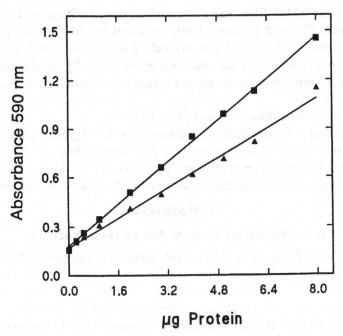

Fig. 1. Coomassie blue staining assay of aliquots of 0.25–8 µL of 1 mg/mL
solutions of bovine serum albumin (■) and γ-globulin (▲).

ber of other chemicals *(1)*. The interfering substances include millimolar concentrations of frequently used buffers such as N-2-hydroxyethylpi-
perazine-N'-2-ethanesulfonic acid, 3-(N-morpholino)propanesulfonic
acid, and tris-(hydroxymethyl)aminomethane and low concentrations
of density gradient materials such as CsCl and metrizamide.

2. It is important to ensure rapid and vigorous mixing of reagent B *(3)*.
3. The standard curve is usually nonlinear. For samples containing low
   concentrations of protein, all absorbance measurements should be made
   at the 750 nm maximum.
4. Filter papers should not be touched with bare hands before the staining
   reaction. It is advisable to leave a border of 1 cm around the filter, as
   the edges may have higher background staining.
5. Low volumes of protein solutions are difficult to dispense accurately
   with air displacement pipets. Greater precision is obtained with positive
   displacement devices such as a 10 µL Hamilton syringe without bevel.
6. There is no interference in the assay of proteins dissolved in the following solutions when fixation (step 2) is performed: N-2-hydroxy-
   ethylpiperazine-N'-2-ethanesulfonic acid (1M), metrizamide (50% [w/

v]), sucrose (2*M*), sodium dodecyl sulfate (1%), tris-(hydroxy-methyl)aminomethane (1*M*), or urea (9*M*). Fixation is not required for most samples, omission of step 2 results in blanks equivalent to <0.5 μg with 8 μL of the most strongly interfering substance from the list above (1% sodium dodecyl sulfate). Very high concentrations of salt cause some inhibition of dye binding (e.g., 5*M* CsCl results in about 20% inhibition).

7. Staining and destaining are performed best with continuous gentle rocking or rotating.

8. Reliable protein estimations may be made of samples containing colored proteins or dyes, such as bromophenol blue or phenol red, which do not bind to proteins under the assay conditions. Thus, samples prepared for polyacrylamide gel electrophoresis dissolved in sodium dodecyl sulfate and mercaptoethanol containing bromophenol blue may be assayed directly. By contrast, in a solution assay such as the Lowry or Bradford assay, samples containing dyes require prior treatment (e.g., protein precipitation followed by redissolution *[1]*); similarly, correction for colored proteins, if they absorb at the assay wavelength, requires a sample blank.

9. The assay allows considerable flexibility of timing: The filters may be left for extended periods (even months) after steps 1, 2, or 5. The eluted dye (step 6) is stable for 1 d (or longer if evaporation of the ethanol is prevented). A rapid estimate of protein concentration can be obtained by direct visual examination immediately after destaining (step 4). If high sensitivity is not needed, the staining period may be reduced (70% of the final staining intensity is obtained in 5 min, step 3).

## References

1. Peterson, G. L. (1983) Determination of total protein. *Methods Enzymol.* **91,** 95–119.
2. Rodriguez-Vico, F., Martinez-Cayuela, M., Garcia-Peregrin, E., and Ramirez, H. (1989) A procedure for eliminating interferences in the Lowry method of protein determination. *Anal. Biochem.* **183,** 275–278.
3. Lowry, O. H., Rosebrough, N. J., Farr, A. L., and Randall, R. J. (1951) Protein measurement with the Folin phenol reagent. *J. Biol. Chem.* **193,** 265–275.
4. Smith, P. K., Krohn, R. I., Hermanson, G. T., Mallia, A. K., Gartner, F. H., Provenzano, M. D., Fujimoto, E. K., Goeke, N. M., Olson, B. J., and Klenk, D. C. (1985) Measurement of protein using bicinchoninic acid. *Anal. Biochem.* **150,** 76–85.
5. Bradford, M. M. (1976) A rapid and sensitive method for the quantitation of microgram quantities of protein utilizing the principle of protein-dye binding. *Anal. Biochem.* **72,** 248–254.

6. Peters, W. H. M., Fleuren-Jakobs, A. M. M., Kamps, K. M. P., de Pont, J. J. H. H. M., and Bonting, S. L. (1982) Lowry protein determination on membrane preparations: Need for standardization by amino acid analysis. *Anal. Biochem.* **124,** 349–352.
7. Markwell, M. A. K., Haas, S. M., Bieber, L. L., and Tolbert, N. E. (1978) A modification of the Lowry procedure to simplify protein determination in membrane and lipoprotein samples. *Anal. Biochem.* **87,** 206–210.

CHAPTER 19

# Separation and Analysis
# of Membrane Proteins
# by SDS-Polyacrylamide
# Gel Electrophoresis

## *Michael J. Dunn and Samantha J. Bradd*

## 1. Introduction

Polyacrylamide gel electrophoresis (PAGE) in the presence of the anionic detergent, sodium dodecyl sulfate (SDS), is probably the most commonly used technique for the analysis of protein mixtures. SDS is a very effective solubilizing agent for a wide range of polypeptides, including membrane proteins. It has been established that the majority of proteins bind 1.4 g SDS/1 g protein (*1*) to form negatively charged complexes. This results in masking of the intrinsic charge of the polypeptide chains, so that net charge per unit mass becomes approximately constant. Although the majority of proteins bind SDS in the expected ratio, it is important to realize that proteins containing nonprotein groups (e.g., glycoproteins, phosphoproteins, lipoproteins) can bind varying amounts of SDS, resulting in anomalous mobility and separation artifacts during SDS-PAGE.

During electrophoresis, separation depends only on the effective mol radius, which roughly approximates to mol size, and occurs as a result of a sieving effect of the gel matrix on the SDS-protein complexes. The polyacrylamide gel concentration used determines the effective separation range of the SDS-PAGE system (Table 1). If the sample to be analyzed contains proteins with a wide range of mol

From: *Methods in Molecular Biology, Vol. 19: Biomembrane Protocols: I. Isolation and Analysis*
Edited by: J. M. Graham and J. A. Higgins  Copyright ©1993 Humana Press Inc., Totowa, NJ

Table 1
Separation Ranges
of Polyacrylamide Gels of Various Concentrations
(%T; *see* Note 1)

| %T | Optimum $M_r$ range |
|------|------------------------|
| 3–5 | >100,000 |
| 5–12 | 20,000–150,000 |
| 10–15 | 10,000–80,000 |
| >15 | <15,000 |

wts, it is then advantageous to use a gradient SDS-PAGE system *(2)* to extend the effective separation range (*see* Chapter 24). For example, a 3–30% T (*see* Note 1) gel can separate proteins in the $M_r$ range 10,000 to nearly 500,000. Small polypeptides and peptides (<15,000) cannot be separated by the standard SDS-PAGE system, as small molecules form SDS-protein complexes of similar dimension and charge, which migrate together during SDS-PAGE. However, separations in the range 1–45 kDa can be achieved using gels with a high *N-N'*-methylene-*bis*-acrylamide (*bis*-acrylamide) crosslinker content (10% C) (*see* Note 1) and containing 8*M* urea, which decreases the size of the detergent micelles *(3,4)*.

SDS-PAGE is a relatively simple and reproducible technique, yet it is capable of very high resolution, as it can separate SDS-protein complexes differing in mobility by as little as 1%, which is equivalent to a difference in $M_r$ of about 1000 for a 100,000 protein. To ensure optimal reaction with SDS, samples should be heated at 100°C for 3–5 min in the presence of 2–5% SDS and reagents such as 0.1*M* β-mercaptoethanol or 20 m*M* dithiothreitol to cleave disulfide bonds.

The most commonly used buffer system for SDS-PAGE is that devised by Laemmli *(5)*, based on the discontinuous buffer system of Ornstein and Davis *(6,7)* with the addition of 0.1% SDS. This buffer system concentrates sample proteins into a narrow starting zone prior to separation resulting in very sharp bands during the separation. This effect is owing to the moving boundary formed between the rapidly moving (leading) and slowly migrating (trailing) species. The lower gel phase contains the leading constituent (chloride), and the upper buffer phase contains the trailing ion (glycinate). As the sample proteins have intermediate mobilities they are concentrated ("stacked") in the moving boundary.

The moving boundary is formed in a low concentration stacking gel, and there is an increase in pH in the more concentrated, restrictive separating gel. This causes further dissociation of glycinate, increasing its mobility so that it moves just behind the chloride ion. This effect, together with the decrease in gel pore size, causes the proteins to become unstacked and to be subsequently separated on the basis of their size.

## 2. Materials

### 2.1. Gel Apparatus

Apparatus suitable for SDS-PAGE is readily available from a number of commercial suppliers (e.g., Bio-Rad Laboratories [Richmond, CA], Bioemtra [Göttingen, Germany], Hoefer Scientific Instruments [San Francisco, CA], Pharmacia Biotechnology [Uppsala, Sweden]). The technique is usually carried out in slab gels, which can be run in a large-format or mini-format vertical gel apparatus.

### 2.2. Stock Solutions

Chemicals should be analytical reagent (Analar) or electrophoresis grade and water should be double distilled. A number of commercial suppliers (e.g., Bio-Rad Laboratories [Poole, UK], Kodak Ltd. [Rochester, NY], Merck BDH, Millipore [Bedford, MA], Pharmacia Biotechnology [Uppsala, Sweden], Sigma [St. Louis, MO]) produce a range of reagents recommended for use in gel electrophoresis.

1. Stock acrylamide solution (30% T, 2.7% C; *see* Note 1): Dissolve 29.2 g acrylamide and 0.8 g *bis*-acrylamide, and make up to 100 mL with water (*see* Note 2). Deionize the solution with 1 g/100 mL Amberlite MB-1 monobed resin for at least 1 h and then filter. The stock solution can be stored in an amber bottle at 4°C for several weeks.
2. Stacking gel buffer ($1M$ Tris-HCl, pH 6.8): Dissolve 121.1 g Tris base in 800 mL distilled water and adjust to pH 6.8 with $1M$ HCl. Make up to a final volume of 1 L. This solution is stable for at least 1 mo at 4°C.
3. Separating gel buffer ($1.5M$ Tris-HCl, pH 8.8): Dissolve 181.65 g Tris base in 800 mL distilled water and adjust to pH 8.8 with $1M$ HCl. Make up to a final volume of 1 L. This solution is stable for at least 1 mo at 4°C.
4. 10% (w/v) ammonium persulfate: Dissolve 1.0 g of ammonium persulfate in 10 mL distilled water. This solution should be prepared daily.
5. 10% (w/v) SDS: Dissolve 1 g SDS and make up to 10 mL with distilled water.

6. 0.1% (w/v) Bromophenol blue: Dissolve 0.1 g bromophenol blue in 100 mL distilled water.
7. Double-strength sample solubilization solution:

| Component | Amount | Final concentration |
|-----------|--------|---------------------|
| 1*M* Tris-HCl, pH 6.8 | 1.25 mL | 0.125*M* |
| 10% SDS | 4.0 mL | 4% |
| 0.1% Bromophenol blue | 0.2 mL | 0.002% |
| Glycerol | 2.0 mL | 20% |
| β-mercaptoethanol | 1.0 mL | 10% |
| Distilled water | 1.55 mL | — |

This recipe is for 10 mL of sample solubilization solution, which can be stored frozen in aliquots.
8. Reservoir buffer (0.192*M* glycine, 0.025*M* Tris, 0.1% SDS): Dissolve 14.4 g glycine, 3.03 g Tris base, and 1 g SDS in distilled water and make up to a final volume of 1 L. This solution is stable for at least 1 mo at 4°C.
9. Fixative: Dissolve 20 g trichloroacetic acid (TCA) and make up to 100 mL with distilled water.
10. *N,N,N',N'*-tetramethylethylenediamine (TEMED): Store at 4°C.
11. Water-saturated isobutanol.
12. Microsyringe or automatic pipet with narrow-ended tips.

## 3. Methods

1. Assemble the gel cassettes using clean, dry glass plates according to the manufacturer's instructions for the particular electrophoresis apparatus being used.
2. Recipes for the preparation of 100 mL of separating gel mixture of different %T are given in Table 2. Mix the stock acrylamide solution, stock 1.5*M* Tris pH 8.8, ammonium persulfate, and water, and then degas using a vacuum pump to remove oxygen, which inhibits polymerization. Then carefully add SDS and TEMED, and mix the solution gently to avoid reoxygenation. Pour the gel solution immediately into the gel cassettes to about 1 cm below the level, which will be occupied by the well-forming combs when they are in position.
3. Overlay the gels carefully with distilled water (or water-saturated isobutanol), and allow to polymerize for at least 2 h (or overnight) at room temperature.
4. Prepare the required volume of stacking gel solution, e.g., 30 mL, using 3.0 mL acrylamide stock, 3.75 mL 1*M* Tris-HCl, pH 6.8, 22.79 mL distilled water, and 135 μL ammonium persulfate. Mix the solution,

Table 2
Recipes for Separating Gel Solutions with Different %T (*see* Note 1)

| Solution | Final concentration of acrylamide (%T) | | | |
| --- | --- | --- | --- | --- |
| | 20 | 15 | 10 | 5 |
| Acrylamide stock | 66.6 mL | 49.95 mL | 33.3 mL | 16.65 mL |
| 1.5*M* Tris-HCl, pH 8.8 | 25.0 mL | 25.0 mL | 25.0 mL | 25.0 mL |
| 10% SDS | 1.0 mL | 1.0 mL | 1.0 mL | 1.0 mL |
| TEMED | 50 µL | 50 µL | 50 µL | 50 µL |
| Ammonium persulfate | 150 µL | 150 µL | 150 µL | 150 µL |
| Distilled water | 7.2 mL | 23.85 mL | 40.5 mL | 57.15 mL |

and degas using a vacuum pump. Then add 3.75 mL SDS and 30 µL TEMED. Mix the solution gently, and use immediately.

5. Pour off the water or isobutanol from the top of the polymerized separating gel, and wash it with water. Fill the cassettes with the stacking gel mixture, and insert the appropriate sample well-forming combs. The gels should be polymerized for at least 2 h at room temperature.

6. When the stacking gels have polymerized, remove the sample combs from the gels, and install the gels in the electrophoresis apparatus according to the manufacturer's instructions.

7. Fill the electrode chambers with the reservoir buffer solution. On some designs of apparatus, the buffer can be circulated between the cathode and anode chambers to equalize pH conditions.

8. Dissolve dry samples in single-strength (i.e., diluted 1:1 with distilled water) sample solubilization solution. Dilute liquid samples to a suitable protein concentration, and add to an equal volume of double-strength sample solubilization solution. The total sample volume should be less than the volume of the sample wells in the gel. Heat the samples in a boiling water bath or block heater at 100°C for 3 min.

9. Load the samples into the sample well using a suitable microsyringe or micropipet fitted with narrow-ended tips.

10. Connect the apparatus to a suitable power pack with the anode connected to the lower reservoir. Run the gels under constant current conditions. 1.5-mm thick, 15-cm long 10%T gels can be run overnight at 15 mA/gel, by which time the bromophenol blue tracking dye should reach the bottom of the gels (*see* Note 3).

11. After electrophoresis, remove the gels from the cassettes, and fix them in a solution of 20% (w/v) TCA prior to visualization of the separated proteins (*see* Notes 4–7). A typical SDS gel separation visualized with Coomassie brilliant blue R-250 is shown in Fig. 1.

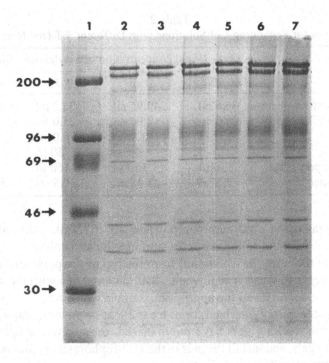

Fig. 1. A 10%T SDS-PAGE separation of human erythrocyte membrane proteins stained with Coomassie brilliant blue R-250. Lane 1 contains colored mol wt markers (Amersham), the arrows indicate the $M_r$ values. Lanes 2–7, erythrocyte membrane proteins, 20 μg *(2,3)*, 25 μg *(4,5)*, and 30 μg *(6,7)* loadings.

## 4. Notes

1. The size of the pores within the gel matrix is dependent both on the total monomer concentration (%T) and on the concentration of crosslinking agent (%C). By convention %T is expressed as total gel concentration (i.e., acrylamide plus *bis*-acrylamide as g/100 mL), whereas %C is the percentage (by weight) of the total monomer that is the crosslinking agent.
2. Acrylamide is known to be neurotoxic. Always dispense acrylamide powder in a fume cupboard and always wear gloves when handling acrylamide powder or solutions. This potential hazard can be minimized by using one of the ready-made forms available from various suppliers. These are either in the form of preweighed powder or stock solutions. A further alternative is provided by a range of ready-made homogeneous or gradient SDS polyacrylamide gels, which are available to fit some designs of commercial apparatus.

3. Higher field strengths can be used to accelerate electrophoresis, but care must be taken to minimize heating effects within the gels. This can be accomplished using a refrigerated recirculating water bath to cool the lower (anode) electrode buffer solution. The mini-gel format can also be used to advantage to provide rapid separations without the need to resort to elevated field strengths.

4. If SDS gels are to be used for Western blotting, the fixation step must be omitted. This technique is described in Chapter 20.

5. For general protein visualization, staining with Coomassie brilliant blue R-250 is recommended *(8)*.

6. For higher sensitivity of detection, silver staining is recommended *(9)*.

7. For the visualization of radioactively labeled proteins, autoradiography or fluorography should be used. The latter method is essential for proteins labeled with low energy β-emitters such as $^3$H, and gives increased sensitivity (i.e., reduced exposure times) for other isotopes. For fluorography, the gels must be impregnated with a suitable scintillant such as PPO *(10)* or an aqueous fluor such as En$^3$Hance (New England Nuclear, Boston, MA), Enlightning (NEN), or Amplify (Amersham International). Gels must be dried under vacuum *(8)* and then exposed to a suitable X-ray film for the appropriate time.

## Acknowledgments

We are grateful to the British Heart Foundation for financial support. We thank Carol Lovegrove for preparing the photograph.

## References

1. Reynolds, J. A. and Tanford, C. (1970) The gross conformation of protein-sodium dodecyl sulfate complexes. *J. Biol. Chem.* **23,** 5161–5165.
2. Walker, J. M. (1984) Gradient SDS polyacrylamide gel electrophoresis, in *Methods in Molecular Biology, vol. 1: Proteins* (Walker, J. M., ed.), Humana, Clifton, NJ, pp. 57–61.
3. Swank, R. T. and Munkries, K. D. (1971) Molecular weight analysis of oligopeptides by electrophoresis in polyacrylamide gel with sodium dodecyl sulfate. *Anal. Biochem.* **39,** 462–477.
4. Anderson, B. L., Berry, R. W., and Telser, A. (1983) A sodium dodecyl sulfate-polyacrylamide gel electrophoresis system that separates peptides and proteins in the molecular weight range of 2,500 to 90,000. *Anal. Biochem.* **132,** 365–375.
5. Laemmli, U. K. (1970) Cleavage of structural proteins during the assembly of the head of bacteriophage T4. *Nature* **227,** 680–685.
6. Ornstein, L. (1964) Disc electrophoresis—I. Background and theory. *Ann. NY Acad. Sci.* **121,** 321–349.

7. Davis, B. J. (1964) Disc electrophoresis—II. Method and application to human serum proteins. *Ann. NY Acad. Sci.* **121,** 404–427.
8. Smith, B. J. (1984) SDS polyacrylamide gel electrophoresis of proteins, in *Methods in Molecular Biology, vol. 1: Proteins* (Walker, J. M., ed.), Humana, Clifton, NJ, pp. 41–55.
9. Patel, K., Easty, D. J., and Dunn, M. J. (1988) Detection of proteins in polyacrylamide gels using an ultrasensitive silver staining technique, in *Methods in Molecular Biology, vol. 3: New Protein Techniques* (Walker, J. M., ed.), Humana, Clifton, NJ, pp. 159–168.
10. Waterborg, J. H. and Matthews, H. R. (1984) Fluorography of polyacrylamide gels containing tritium, in *Methods in Molecular Biology, vol. 1: Proteins* (Walker, J. M., ed.), Humana, Clifton, NJ, pp. 147–152.

CHAPTER 20

# Analysis of Membrane Proteins by Western Blotting and Enhanced Chemiluminescence

*Samantha J. Bradd and Michael J. Dunn*

## 1. Introduction

### 1.1. Background

The high resolution capacity of polyacrylamide gel electrophoresis (PAGE) (*see* Chapter 19) has resulted in the widespread application of this group of techniques to protein separations. PAGE procedures can provide characterization of proteins in terms of their charge, size, relative hydrophobicity, and abundance. However, they provide no direct clues as to the identity or function of the separated components. A powerful approach to this problem is provided by probing the separated proteins with antibodies and other ligands specific for components of the protein mixture being analyzed.

Early attempts to exploit this approach involved the reaction of gels after electrophoresis with solutions of the probes being employed, this technique being termed immunofixation. However, the polyacrylamide gel matrix is restrictive to the diffusion of probe molecules, so that prolonged incubation and washing steps have to be used to achieve a strong specific signal with a low background. Unfortunately, the separated protein zones are also subject to diffusion during this time resulting in loss of resolution and merging of adjacent bands.

These problems were solved with the advent of protein-blotting techniques, based on those developed by Southern (*1*) for analysis of electrophoretic separations of DNA. In these procedures the proteins,

From: *Methods in Molecular Biology, Vol. 19: Biomembrane Protocols: I. Isolation and Analysis*
Edited by: J. M. Graham and J. A. Higgins Copyright ©1993 Humana Press Inc., Totowa, NJ

after separation by electrophoresis, are transferred ("blotted") onto the surface of a membrane such as nitrocellulose. When immobilized on the surface of a membrane, the proteins are readily accessible to interaction with probes. Moreover, relatively short incubation times can be used, and the proteins themselves are no longer subject to diffusion.

## *1.2. Experimental Strategy*

### *1.2.1. Apparatus*

Various techniques have been employed for the transfer of electrophoretically separated proteins to membranes. However, the most popular method is the application of an electric field perpendicular to the plane of the gel. This technique of electrophoretic transfer, first described by Towbin et al. *(2)*, is known as Western blotting. Two types of apparatus are in routine use for electroblotting. In the first approach the sandwich assembly of gel and blotting membrane is placed vertically between two platinum wire electrode arrays contained in a tank filled with blotting buffer *(3)*. The disadvantages of this technique are that (1) a large volume of blotting buffer must be used, (2) efficient cooling must be provided if high current settings are employed to facilitate rapid transfer, and (3) the field strength applied (V/cm) is limited by the relatively large interelectrode distance. In the second type of procedure the gel-blotting membrane assembly is sandwiched between two horizontal plate electrodes, typically made of graphite *(4)*. The advantages of this method are that (1) relatively small volumes of transfer buffer are used, (2) special cooling is not usually required, although the apparatus can be run in a cold room if necessary, and (3) a relatively high field strength (V/cm) can be applied owing to the short interelectrode distance resulting in faster transfer times.

### *1.2.2. Blotting Membranes*

A variety of membranes have been described for Western blotting *(3)*, but nitrocellulose is still the most popular. Supported nitrocellulose membranes that overcome the problem of the fragility of this matrix are now available. Recently other membranes based on the use of polyvinylidine difluoride (PVDF) or polypropylene have been produced. These membranes have the advantages of robustness and high protein-binding capacity. These membranes work well in immunoblotting protocols *(5)* and are used extensively in techniques that allow the direct chemical characterization of proteins separated by PAGE (e.g., microsequencing, amino acid analysis, peptide mapping) *(6)*.

### 1.2.3. Visualization

After transfer, total protein patterns can be visualized by total protein staining, usually with Amido black *(3)*. Before reaction with specific ligands, blots must be reacted with a reagent to block any protein-binding sites remaining on the membrane. A variety of reagents such as bovine serum albumin, gelatin, and nonionic detergents can be used for this blocking step *(3)*, and more recently dried nonfat milk has become popular for this purpose *(7)*. After blocking, the blot is reacted with the specific ligand. Visualization of the bound ligand can be achieved by labeling of the primary ligand itself, e.g., by radiolabeling (usually $^{125}$I) or enzyme conjugation. However, this approach is not generally popular, as derivitization of the primary ligand can often interfere with its reactivity. Generally an indirect technique is used in which the blot is reacted with a second reagent specific for the primary ligand. For antibodies this involves the use of an anti-Ig specific for the species in which the primary antibody was produced. Again visualization can be achieved by radiolabeling, e.g., $^{125}$I-protein A *(8)*. However, techniques in which the secondary reagent is conjugated with an enzyme (e.g., horseradish peroxidase, alkaline phosphatase) and subsequently visualized with a colored reaction product are much more popular *(9)*. The sensitivity of these methods can be substantially increased using the avidin-biotin system *(10)* or colloidal gold techniques *(11)*.

More recently, visualization techniques based on the use of chemiluminescent reactions of reagents such as cyclic diacylhydrazides have been developed for use with Western blots *(12)*. This chapter describes the use of one such method in which horseradish peroxidase is used to catalyze the oxidation of luminol in the presence of hydrogen peroxide under alkaline conditions. This reaction results in the emission of light and can be increased up to 1000-fold by certain enhancers (e.g., para-substituted phenolic compounds). This enhanced chemiluminescent (ECL) system is used in conjunction with standard immunodetection protocols employing peroxidase-conjugated secondary antibodies as shown in Fig. 1. The image is visualized by expo-

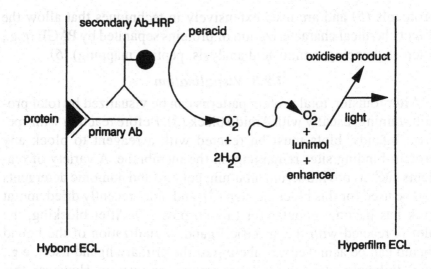

Fig. 1. Diagram showing the mechanism of detection of proteins on Western blots using the ECL system.

sure of the blot to an appropriate blue-sensitive film, usually for times of between 15 s and 1 h. This method provides at least a 10-fold increase in sensitivity over other detection methods, and facilitates the detection of <1 pg of protein.

## 2. Materials

### 2.1. Semidry Blotting

Prepare buffers from analytical grade reagents and dissolve in deionized water. Buffers should be stored at 4°C and are stable for up to 3 mo.

1. Transfer/equilibration buffer: 20 m$M$ Tris, 150 m$M$ glycine, pH 8.3. Dissolve 2.42 g Tris base and 11.26 g glycine and make up to 1 L. The pH of the buffer is pH 8.3 and should not require adjustment.
2. Transfer membrane: Hybond-ECL nitrocellulose transfer membrane (Amersham International) cut to the size of the gel to be blotted (*see* Note 1).
3. Filter paper: Whatman 3MM filter paper cut to the size of the gel to be blotted.
4. Equipment: A number of commercial companies produce semidry electroblotting apparatus and associated power supplies. We have used the NovaBlot apparatus from Pharmacia Biotechnology.

5. Rocking platform.
6. Plastic boxes for gel incubations.

## 2.2. Protein Detection
## Using Enhanced Chemiluminescence (ECL)

1. Washing solution (PBS-T): Phosphate buffered saline (PBS) containing 0.05% (w/v) Tween 20. Make 0.05 g Tween 20 up to 100 mL with PBS (8.0 g/L NaCl, 0.2 g/L KCl, 1.15 g/L $Na_2HPO_4 \cdot 2H_2O$, 0.2 g/L $KH_2PO_4$).
2. Blocking solution (PBS-TM): PBS-T with the addition of 3% (w/v) nonfat dried milk (Marvel). Dissolve 3 g milk powder and make up to 100 mL with PBS-T.
3. Secondary antibody reagent: Peroxidase-conjugated rabbit immunoglobulins to mouse immunoglobulins (DAKO, Carpinteria, CA) diluted 1:1000 in PBS-TM (*see* Note 2).
4. ECL Western blotting detection kit (Amersham International).
5. Detection film: Hyperfilm ECL (Amersham International) (*see* Note 3).
6. Diaminobenzidine (DAB) solution: Contains 0.05% (w/v) DAB and 0.01% (v/v) $H_2O_2$. Dissolve 0.05 g DAB in 100 mL of PBS, and add 0.01 mL $H_2O_2$ (30% stock solution).
7. Photographic equipment: Photoradiography cassettes are required for exposure of the blots with ECL film. The film can be developed using an automatic X-ray film developing unit or by hand using standard X-ray developer and fixer.
8. Rocking platform.
9. Plastic bags for incubations.
10. Glass plate.
11. Saran Wrap™.

## 3. Methods
### 3.1. Semidry Blotting

This technique has been described in detail in vol. 3 of this series *(4)* and will only be briefly described below.

1. Following separation of the membrane proteins by SDS-PAGE (*see* Chapter 19), place the gel in equilibration buffer, and gently agitate on a rocking platform for 30 min at room temperature.
2. Stack six sheets of filter paper wetted with blotting buffer on the lower (anode) plate of the electroblotter, followed by the prewetted nitrocellulose transfer membrane. Then carefully place the gel on top of the membrane. Apply a further six sheets of wetted filter paper, and place the upper plate (cathode) in position (*see* Note 4).
3. Transfer the proteins at 0.8 mA/cm$^2$ of gel for 40 min (*see* Note 5).

### *3.2. Protein Detection by ECL*

Carry out all incubations with gentle agitation on a rocking platform at room temperature.

1. Following transfer, gently remove the transfer membrane from the electroblotter; seal in a plastic bag containing the blocking solution, and incubate for 1 h.
2. Remove the blocking solution and incubate the blot for 1 h with the primary antibody diluted at an appropriate concentration in PBS-TM (*see* Note 6). In the example illustrated in Fig. 2 (strips 2 and 3), mouse monoclonal antibodies to the $\alpha$- and $\beta$-subunits of erythrocyte spectrin were diluted 1:10,000 and 1:1000 respectively for visualization using ECL.
3. Wash the blot three times for 5 min with PBS-T.
4. Incubate the blot for 1 h with peroxidase-conjugated secondary antibody diluted at an appropriate concentration (1:1000 in this example) in PBS-TM.
5. Wash the blot three times for 5 min in PBS alone.
6. Mix the two ECL reagents in a ratio of 1:1 to give a final volume of $0.125 \text{ mL/cm}^2$ transfer membrane. Incubate the blot in the ECL reagent mixture for 1 min.
7. Drain excess solution from the blot; then place on a clean glass plate (protein side uppermost) and cover with food wrapping film (e.g., Saran Wrap™). It is important that the blot should not be allowed to dry out at this stage. Air bubbles and creases in the film must be avoided.
8. Expose the blot to the film in the dark for 15 s to 60 min and then develop using an automated processor or by hand. In the example illustrated in Fig. 2 an exposure of 1 min was used.
9. After completion of the ECL exposure, the transfer membrane can be washed (2 × 15 min in PBS-T) and visualized by the addition of DAB solution for 10 min at room temperature.

## 4. Notes

1. Hybond ECL is a 0.45-μm pore size nitrocellulose membrane. Although this is the recommended immobilizing matrix to be used with the ECL detection kit, other nitrocellulose membranes and nylon filters are also suitable.
2. The appropriate peroxidase-conjugated anti-Ig specific for the species in which the primary antibody was produced must be used.
3. The film used to detect the signal produced by the ECL system must be sensitive to blue light. Hyperfilm ECL has been specifically chosen to give optimum results with the ECL technique.

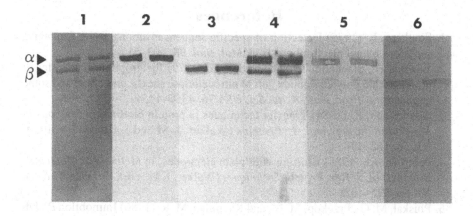

Fig. 2. Result of an immunoblotting experiment to detect the α- and β-subunits of spectrin after SDS-PAGE separation of human erythrocyte membrane proteins. 1, Amido Black stain of blot; 2 to 4, visualization using the ECL system; 5 and 6, visualization using DAB. Monoclonal antibodies to α-spectrin (2, diluted 1:10,000; 5, diluted 1:1000), β-spectrin (3, diluted 1:1000; 6, diluted 1:200), αβ-spectrin (4, diluted 1:500).

4. The method for setting up the semidry blotting apparatus is generally applicable and is appropriate for the equipment that we have used. Instructions provided with equipment from other manufacturers should be followed.

5. The maximum mA/cm$^2$ of gel quoted applies to the apparatus we have used. This should be established from the manual for the particular equipment available.

6. The appropriate dilutions of the primary and secondary antibody reagents and the exposure time required to obtain optimal results will vary for each particular primary antibody. These parameters must be established using serial dilutions and varying exposure times so that a strong specific signal is obtained with minimal nonspecific background staining. For visualization using ECL, the antibodies can usually be diluted 5–10 times more than for visualization using DAB.

## Acknowledgments

We would like to thank the British Heart Foundation for financial support. We are grateful to Julie Guilford, Amersham International, for help in the establishment of the ECL technique in our laboratory and to Carol Lovegrove for preparing the photographs.

# References

1. Southern, E. (1975) Detection of specific sequences among DNA fragments separated by gel electrophoresis. *J. Mol. Biol.* **98,** 503–517.
2. Towbin, H., Staehelin, T., and Gordon, G. (1979) Electrophoretic transfer of proteins from polyacrylamide gels to nitrocellulose sheets: procedure and some applications. *Proc. Natl. Acad. Sci. USA* **76,** 4350–4354.
3. Gooderham, K. (1984) Transfer techniques in protein blotting, in *Methods in Molecular Biology, vol. 1: Proteins* (Walker, J. M., ed.), Humana, Clifton, NJ, pp. 165–178.
4. Peferoen, M. (1988) Blotting with plate electrodes, in *Methods in Molecular Biology, vol. 3: New Protein Techniques* (Walker, J. M., ed.), Humana, Clifton, NJ, pp. 395–402.
5. Pluskal, M. G., Przekop, M. B., and Kavonian, M. R. (1986) Immobilon PVDF transfer membrane: a new membrane substrate for Western blotting of proteins. *Bio Techniques* **4,** 272–283.
6. Aebersold, R. (1991) High sensitivity sequence analysis of proteins separated by polyacrylamide gel electrophoresis, in *Advances in Electrophoresis, vol. 4* (Chrambach, A., Dunn, M. J., and Radola, B. J., eds.), VCH, Weinheim, pp. 81–168.
7. Jagus, R. and Pollard, J. W. (1988) Use of dried milk for immunoblotting, in *Methods in Molecular Biology, vol. 3: New Protein Techniques* (Walker, J. M., ed.), Humana, Clifton, NJ, pp. 403–408.
8. Kruger, N. J. and Hammond, J. B. W. (1988) Immunodetection of proteins on "Western" blots using [125]I-labelled protein A, in *Methods in Molecular Biology, vol. 3: New Protein Techniques* (Walker, J. M., ed.), Humana, Clifton, NJ, pp. 408–417.
9. Walker, J. M. and Gaastra, W. (1988) Detection of protein blots using enzyme-linked second antibodies or protein A, in *Methods in Molecular Biology, vol. 3: New Protein Techniques* (Walker, J. M., ed.), Humana, Clifton, NJ, pp. 427–440.
10. Dunn, M. J. and Patel, K. (1988) Detection of protein blots using the avidin-biotin system, in *Methods in Molecular Biology, vol. 3: New Protein Techniques* (Walker, J. M., ed.), Humana, Clifton, NJ, pp. 419–426.
11. Jones, A. and Moeremans, M. (1988) Colloidal gold for the detection of proteins on blots and immunoblots, in *Methods in Molecular Biology, vol. 3: New Protein Techniques* (Walker, J. M., ed.), Humana, Clifton, NJ, pp. 441–479.
12. Durrant, I. (1990) Light-based detection of biomolecules. *Nature* **346,** 297–298.

# CHAPTER 21

# Two-Dimensional Polyacrylamide Gel Electrophoresis of Membrane Proteins

## Joe Corbett and Michael J. Dunn

## 1. Introduction

Two-dimensional polyacrylamide gel electrophoresis (2-D PAGE) using isoelectric focusing (IEF) in the first dimension, followed by sodium dodecyl sulfate polyacrylamide gel electrophoresis (SDS-PAGE) in the second as described by O'Farrell (1), has been widely used for almost 20 years in examining complex protein mixtures (2). During this time a large variety of modifications to the original technique have been employed in efforts to obtain more reproducible results with greater resolution and protein spot detection, or to visualize specific populations of polypeptides.

The replacement of nonionic detergents (Nonidet P-40 or Triton X-100) with zwitterionic detergents such as 3-([3-cholamidopropyl]-dimethylammonio)-1-propanesulfonate (CHAPS) in the first dimension separation procedure is one such modification (3–5). Many variations in gel composition in both the first and second dimensions have also been suggested (6,7). Different running conditions in terms of buffer concentrations and voltage settings have been utilized with some workers replacing prefocusing of IEF gels with stepwise voltage increases during the run (7,8). Synthetic carrier ampholytes are now commercially available from several sources, and these products differ considerably in pH gradient formation properties (9). Flatbed

From: *Methods in Molecular Biology, Vol. 19: Biomembrane Protocols: I. Isolation and Analysis*
Edited by: J. M. Graham and J. A. Higgins Copyright ©1993 Humana Press Inc., Totowa, NJ

IEF procedures have become increasingly popular in recent years, particularly with the development of immobilized pH gradient gels (IPGs), which eliminate the problem of cathodic drift *(10–12)*. Sample preparation is another aspect of the 2-D PAGE procedure that can give rise to great protein pattern variation and lead to difficulties in the interpretation and comparison of results from different laboratories.

The need for a standard 2-D PAGE protocol allowing the reproducible production of good quality gels from IEF and nonequilibrium pH gradient electrophoresis (NEPHGE) *(13)*, which may be compared between laboratories, is obvious. Although the Anderson ISO-DALT® system *(9)* provides such a method for large scale 2-D gel work, this system is unlikely to be available to the majority of researchers working in laboratories not specializing in 2-D electrophoresis. The aim of this chapter is to detail a procedure that has been found to yield consistently good separations of a variety of samples in our laboratory.

## 2. Materials

### 2.1. Gel Apparatus

Apparatus for running first-dimension IEF gels in glass tubes is available from a number of commercial suppliers (e.g., Bio-Rad [Richmond, CA], Biometra [Gottingen, Germany], Hoefer Scientific Instruments [San Francisco, CA], Pharmacia LKB Biotechnology [Uppsala, Sweden]). The apparatus for running the second-dimension gels is identical to that used for one-dimensional SDS-PAGE (*see* Chapter 19). In addition, specialized equipment for running a large number of 2-D separations simultaneously is available from Hoefer Scientific Instruments and Millipore (Bedford, MA).

### 2.2. Stock Solutions

All chemicals should be analytical reagent (Analar) or electrophoresis grade, and water should be double-distilled. A number of commercial suppliers (e.g., Bio-Rad [Richmond, CA], Kodak [Rochester, NY], Merck BDH [Poole, UK], Millipore [Bedford, MA], Pharmacia LKB Biotechnology [Uppsala, Sweden], Sigma [St. Louis, MO]) produce a range of reagents recommended for use in IEF and SDS-PAGE. For IEF gels we use Resolyte 4-8 (Merck), and for NEPHGE gels we use Resolyte 3-10 and Pharmalyte 8-10.5 (Pharmacia).

1. Stock buffer solutions (store at 4°C): Tris-HCl buffer, pH 8.8 (1.5$M$) and pH 6.8 (1$M$).
2. 10% (w/v) SDS stock solution: Store at room temperature.
3. Acrylamide stock solution (30.8%T, 2.6%C): Dissolve 300 g acrylamide and 8 g *bis*-acrylamide in 800 mL distilled water and deionize with Amberlite MB-1 resin (1 g/100 mL) for 1 h, filter and make up to 1 L final vol. Store in the dark at 4°C (*see* Chapter 19).
4. Sample lysis solutions: For IEF gels, mix 5.4 g urea, 6.5 mL double-distilled water, 0.4 g CHAPS, 0.1 g dithiothreitol, 0.5 mL BDH Resolyte 4-8, and 500 μL bromophenol blue (0.05% [w/v]). For NEPHGE gels, mix 5.4 g urea, 6.5 mL double-distilled water, 0.4 g CHAPS, 270 μL β-mercaptoethanol, 375 μL BDH Resolyte 3-10, 125 μL Pharmalyte 8-10.5, and 500 μL bromophenol blue (0.05% [w/v]). For both types of gel, first dissolve the urea in distilled water and deionize with 0.2 g Amberlite MB-1 resin for 1 h to remove any cyanate ions resulting from urea degradation, which can cause protein carbamylation. Then filter the solution before the addition of the remaining ingredients. Aliquot the solution and store at –70°C.
5. Electrode buffers: 20 m$M$ sodium hydroxide (catholyte); 10 m$M$ phosphoric acid (anolyte).
6. Laemmli sample solubilization buffer: Mix 3.75 mL SDS (10% [w/v]), 1.87 mL 1$M$ Tris-HCl buffer (pH 6.8), 625 μL β-mercaptoethanol, 3.75 mL glycerol (75% [w/v]), 1.75 mL double-distilled water, and 350 μL bromophenol blue (0.05% [w/v]). Make up fresh before use.
7. Equilibration buffer: 0.5 mL 1$M$ Tris-HCl (pH 6.8), 2.0 mL SDS (10% [w/v]), 3.8 mL double-distilled water, 0.032 g dithiothreitol, and 250 μL bromophenol blue (0.05% [w/v]).
8. Ammonium persulfate.
9. Hydrochloric acid (conc.).
10. 4$M$ urea (for NEPHGE gels).
11. Water-saturated isobutanol.
12. $N,N,N',N'$-tetramethylethylenediamine (TEMED).
13. Methanol:acetic acid:distilled $H_2O$ (5:1:4 [v/v/v]).

## *2.3. Ancillary Equipment*

1. Microcentrifuge with 1.5-mL tubes.
2. Graduated cylinder (50 mL).
3. Scintered glass filter.
4. Vacuum pump and Buchner flask for degassing.
5. Syringe (10–20 mL).

6. Hamilton syringe (100 μL).
7. Parafilm.

# 3. Methods
## 3.1. Sample Preparation

1. Homogenize the sample in 1% SDS and spin in a microcentrifuge for 5 min; harvest the supernatant and determine the protein concentration using the Bradford assay procedure *(14)* *(see also* Chapter 18).
2. Store samples at –70°C.
3. Once the IEF apparatus has been set up according to the manufacturer's instructions, thaw the samples and mix with at least twice their volume of lysis buffer and load immediately. If samples are initially dissolved in a lysis buffer containing urea, then repeated thawing and refreezing may cause protein carbamylation and result in the appearance of artifactual charge trains on the final 2-D gel pattern *(see* Note 1).

## 3.2. Casting of IEF Rod Gels

1. Clean glass tubes (20 cm in length, 1.5 mm id) in conc. hydrochloric acid; rinse thoroughly in distilled deionized water and acetone before drying in a hot air cabinet.
2. Mark the required gel length (14.5 cm) on the tubes with a felt-tip pen and then place in a clean graduated cylinder.
3. Make 20 mL of IEF gel mixture as follows: 10 g urea, 7.4 mL double-distilled water, and 3.0 mL acrylamide stock solution. Dissolve the urea and deionize by stirring with 0.2 g Amberlite MB-1 resin for 1 h. Filter and degas the mixture before the addition of 0.3 g CHAPS, 1 mL Resolyte 4-8, 15 μL TEMED, and 30 μL 10% (w/v) ammonium persulfate (freshly prepared).
4. Pour the gel mixture into the graduated cylinder containing the tubes until the acrylamide rises to the level marked. The top 0.5 cm of gel solution does not polymerize in contact with the air and serves as an overlay.
5. Allow the gels to polymerize for 1.5 h.
6. Remove tubes from the cylinder and trim off excess acrylamide.

## 3.3. Isoelectric Focusing

1. Insert the rod gels in the top chamber of the electrophoresis tank.
2. Fill the lower anode chamber with 10 m$M$ phosphoric acid, and flush any trapped air bubbles from the ends of the tubes using a buffer-filled syringe.
3. Degas the cathode buffer (20 m$M$ sodium hydroxide) to remove $CO_2$ and add to the top chamber. Remove air remaining in the upper part of

the tubes using a Hamilton syringe filled with cathode buffer ensuring that the unpolymerized urea solution is not disturbed.

4. Apply samples (20–50 µL) using a Hamilton syringe (*see* Note 2).
5. Carry out isoelectric focusing at a constant setting of 800 V for 20 h (i.e., 16,000 volthours) (*see* Note 3).
6. Extrude gels onto labeled lengths of Parafilm using a water-filled syringe attached to the tube with a short piece of rubber tubing (*see* Note 4) and store at –70°C.

### 3.4. NEPHGE

1. Cast the gels as before, substituting 660 µL Resolyte 3-10 and 330 µL Pharmalyte 8-10.5 for the Resolyte 4-8 used in IEF gels.
2. Fill the lower chamber with degassed 20 m$M$ sodium hydroxide; insert the tubes and flush trapped air bubbles from the ends of the tube gels using a syringe.
3. Dissolve the samples in NEPHGE lysis buffer, load and overlay with 10 µL 4$M$ urea before addition of the 10 m$M$ phosphoric acid anode buffer to the top chamber.
4. Connect the upper chamber to the anode and the lower to the cathode before focusing at 600 V overnight (10,000 volthours total) (*see* Note 5).

### 3.5. Second-Dimension Electrophoresis

1. The second-dimension SDS-PAGE procedure is essentially that described by Laemmli *(15)*. Cast the 10%T SDS-polyacrylamide gels as described in Chapter 19 with one modification: Overlay the stacking gels with water-saturated isobutanol in place of the "sample combs" used in casting gels for conventional 1-D SDS-PAGE.
2. Place the IEF rod gels directly onto the second-dimension gel cassette while still frozen (*see* Section 3.3.) and add 1 mL of equilibration buffer. This serves to lubricate the rod gel so that it falls smoothly into place on top of the second-dimension gel. Ensure that good contact exists between the first- and second-dimension gels; remove any air bubbles using a scalpel blade (*see* Note 6). Equilibrate gels *in situ* for 5 min, then drain off the equilibration buffer by turning each cassette on its side. At this stage mol-wt standards can be applied to the second-dimension gels (*see* Note 7).
3. Run gels at 60 mA constant current.
4. After electrophoresis fix the gels in a methanol:acetic acid:distilled water solution (5:1:4 [v/v/v]) or in a 10% (w/v) trichloroacetic acid solution before visualization of the protein pattern using an appropriate proce-

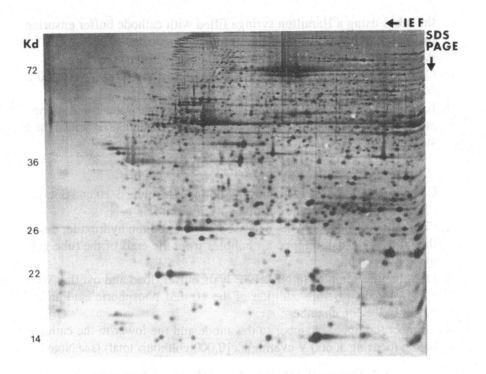

Fig. 1. Typical separation profile obtained using the 2-D PAGE method described. The sample contained 15 µg of human total myocardial proteins. The pattern was visualized by silver staining. The protein mol wts, indicated on the left, were calculated using standard marker proteins.

dure (e.g., staining with Coomassie brilliant blue R-250, silver staining, autoradiography) (*see* Note 2). A typical 2-D pattern produced using this procedure and visualized by silver staining is shown in Fig. 1.

## 4. Notes

1. Proteins can react with cyanate ions produced by the degradation of urea. This results in multiple, carbamylated forms of the protein, each differing by the number of negative charges that have been added. These forms have the same mol wt, so that on the final 2-D pattern they appear as a series (train) of spots of increasing negative charge.
2. The amount of protein that is loaded on an IEF gel depends on the protein visualization methods to be employed and the relative abundance of the proteins of interest in terms of the total sample protein content. We normally load 15 µg of protein for silver staining and 100 µg for Coomassie blue staining.

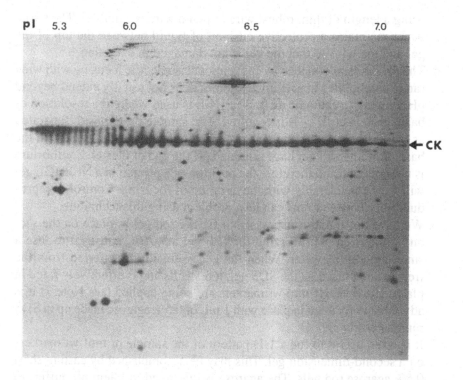

pI   5.3        6.0        6.5        7.0

←CK

Fig. 2. 2-D separation showing carbamylated creatine kinase pI markers visualized by silver staining. The sample also contained a small amount of human myocardial proteins to give reference points for the comparison and matching of other 2-D patterns. The relative pI values indicated were estimated from the known isoelectric point values of the charge train proteins.

3. Before proceeding with a complete study it is advisable to optimize the IEF running conditions and assess the pH gradient produced by a particular batch of ampholytes. This is conveniently achieved by including a series of commercially available pI markers in the sample mixture. These markers are produced by the carbamylation of a purified preparation of a standard protein such as creatine kinase. When applied to the linear pH gradient the markers appear as a row of equidistant spots (Fig. 2). Thus any gaps in the pH gradient can readily be seen. A timecourse can be performed to determine the conditions giving the best resolution of protein spots across the entire length of the IEF gel with minimum loss of basic proteins owing to cathodic drift.

4. As an alternative to the water-filled syringe method of gel extrusion, rod gels can also be removed by inserting a small, tightly fitting piece of tissue paper into the top of the glass tube and pushing the gel out

using a length of thin, robust wire (a piano wire is suitable). The tissue paper maintains a protective cushion of liquid between the top of the gel and the wire so that the gel is not damaged in extrusion.

5. The pH gradient obtained by conventional isoelectric focusing with wide range ampholytes in gels in glass tubes does not usually extend beyond pH 7 owing to cathodic drift. NEPHGE is used where the resolution of basic proteins (pH > 7) is required. The sample proteins are separated as cations and electrophoresis is carried out for short times so that the basic proteins will not have migrated off the end of the gel. Equilibrium is not obtained and therefore the position of a protein in a NEPHGE gel will vary considerably with the duration of focusing. Control of reproducibility between runs can be a problem using this technique.

6. We have not found it necessary to fix the rod gel in place on the second-dimension gel. Freezing the IEF gel at −70°C causes it to lose a small amount of water. When the gel is quickly transferred from the freezer and placed on the SDS gel, it reswells and is effectively fixed in place. However, if mol-wt markers are being applied (*see* Note 7) it is advisable to fix them in place with 1 mL of 0.5% agarose made up in SDS running buffer.

7. It is often useful to run a 1-D pattern of the sample or mol-wt markers on a second-dimension gel. This may be accomplished by casting short 0.5% agarose rod gels. The agarose is dissolved in Laemmli buffer, to which is added the required volume of sample or marker proteins. This mixture is heated to 100°C, pipeted into a 1.5 mm id glass tube and allowed to set. After gel extrusion, lengths of 1 cm can then be cut and applied to the second-dimension gel.

## Acknowledgments

We are grateful to the British Heart Foundation for financial support. We would like to thank Carol Lovegrove for preparing the photographic illustrations.

## References

1. O'Farrell, P. H. (1975) High resolution two-dimensional electrophoresis of proteins. *J. Biol. Chem.* **250,** 4007–4021.
2. Dunn, M. J. (1987) Two-dimensional polyacrylamide gel electrophoresis, in *Advances in Electrophoresis,* vol. 1 (Chrambach, A., Dunn, M. J., and Radola, B. J., eds.), VCH, Weinheim, pp. 1–109.
3. Perdew, G. H., Schaup, H. W., and Selivonchick, D. P. (1983) The use of a zwitterionic detergent in two-dimensional electrophoresis of trout liver microsomes. *Anal. Biochem.* **135,** 453–455.

4. Hjelmeland, L. M. and Chrambach, A. (1981) Electrophoresis and electrofocusing in detergent containing media: A discussion of the basic concepts. *Electrophoresis* **2**, 1–11.
5. Rabilloud, T., Gianazza, E., Catto, N., and Righetti, P. G. (1990) Amidosulphobetaines, a family of detergents with improved solubilization properties: Applications for isoelectric focusing under denaturing conditions. *Anal. Biochem.* **185**, 94–102.
6. Hochstrasser, D. F., Harrington, M. G., Hochstrasser, A. C., Miller, M. J., and Merrill, C. M. (1988) Methods for increasing the resolution of two-dimensional protein electrophoresis. *Anal. Biochem.* **173**, 424–435,
7. Das, J. and Baese, H. J. (1987) Fast method for two-dimensional electrophoresis of proteins from biological samples. *Anal. Biochem.* **164**, 175–180.
8. Duncan, R. and Hershey, J. W. B. (1984) Evaluation of isoelectric focusing running conditions during two-dimensional isoelectric focusing/sodium dodecyl sulphate-polyacrylamide gel electrophoresis: Variation of gel patterns with changing conditions and optimised isoelectric focusing conditions. *Anal. Biochem.* **138**, 144–155.
9. Anderson, L., Hoffman, J. P., Anderson, E., Walker, B., and Anderson, N. G. (1989) Technical improvements in 2-D gel quality and reproducibility using the ISO-DALT system, in *Two-Dimensional Electrophoresis: Proceedings of the International Two-Dimensional Electrophoresis Conference, Vienna* (Endler, A. T. and Hanash, S., eds.), VCH, Weinheim, pp. 288–297.
10. Dunn, M. J. and Patel, K. (1988) Two-dimensional polyacrylamide gel electrophoresis using flat-bed isoelectric focusing in the first dimension, in *Methods in Molecular Biology, Vol. 3: New Protein Techniques* (Walker, J. M., ed.), Humana, Clifton, NJ, pp. 217–231.
11. Righetti, P. G. and Gelfi, C. (1988) Two-dimensional polyacrylamide gel electrophoresis using flat-bed isoelectric focusing in the first dimension, in *Methods in Molecular Biology, Vol. 3: New Protein Techniques* (Walker, J. M., ed.), Humana, Clifton, NJ, pp. 233–256.
12. Gorg, A., Postel, W., and Gunther, S. (1988) The current state of two-dimensional electrophoresis with immobilized pH gradients. *Electrophoresis* **9**, 531–546.
13. O'Farrell, P. Z., Goodman, H. M., and O'Farrell, P. H. (1977) High resolution two-dimensional electrophoresis of basic as well as acidic proteins. *Cell* **12**, 1133–1142.
14. Bradford, M. M. (1976) A rapid and sensitive method for the quantitation of microgram quantities of protein utilizing the principle of protein-dye binding. *Anal. Biochem.* **72**, 248–254.
15. Laemmli, U. K. (1970) Cleavage of structural proteins during the assembly of the head of bacteriophage T4. *Nature* **227**, 680–685.

# Chapter 22

# Isolation
# of a Membrane-Bound Enzyme,
# 5'-Nucleotidase

## *J. Paul Luzio and Elaine M. Bailyes*

## 1. Introduction
### *1.1. Background*

Membrane enzymes may be peripheral or integral proteins. The attachment of peripheral enzymes is mediated mainly by association with other membrane proteins and consists primarily of hydrophilic, electrostatic interactions. Integral membrane proteins are primarily bound by hydrophobic interactions with the lipid bilayer, either via one or more transmembrane peptide domains *(1)* or by possession of a glycosyl-phosphatidylinositol (GPI) membrane anchor at the carboxy-terminus *(2)*. The purification of a membrane enzyme thus requires two additional considerations compared to the isolation of a cytosolic enzyme: the isolation of a membrane fraction containing the enzyme and the solubilization of the enzyme from this fraction. Once solubilization has been achieved and maintained, the enzyme can be treated essentially as a soluble enzyme and subjected to a wide range of purification techniques.

A suitable purification scheme for a membrane enzyme is likely to be based on the following framework:

1. Homogenization of tissue and isolation of a membrane fraction by differential centrifugation.
2. Solubilization.

From: *Methods in Molecular Biology, Vol. 19: Biomembrane Protocols: I. Isolation and Analysis*
Edited by: J. M. Graham and J. A. Higgins  Copyright ©1993 Humana Press Inc., Totowa, NJ

3. Purification steps based on general protein properties such as size, charge, and lectin binding.
4. Affinity purification specific for the enzyme.

Each of these steps is discussed in Section 1.2. Sections 2. and 3. detail a protocol for the purification of rat liver 5'-nucleotidase *(3)*, a widely distributed plasma membrane ectoenzyme *(4,5)* of known primary sequence *(6)*, which is an integral membrane protein attached to the membrane by a GPI anchor *(3,7)*.

## 1.2. Experimental Strategy

### 1.2.1. Homogenization

The method of homogenization depends on the nature and quantity of the tissue used. Solid tissues such as rat liver are readily homogenized using a motor-driven glass/Teflon Potter-Elvehjem homogenizer with a tissue to buffer ratio of 1:5. Homogenizers with a capacity as small as 1 mL total vol can be obtained. The presence of fibrous proteins such as myosin and collagen makes homogenization more difficult. Skeletal muscle, for example, should be chopped up or minced before homogenization in a Waring blender (Fisons, Loughborough, UK). A device with counter-rotating blade (Polytron, Kinematica GmbH, Luzern, Switzerland) can also be effective. Large quantities of tissue, of approx 150–200 g, are most easily processed in a Waring blender.

Homogenization of cells in suspension and cultured cells presents some difficulty. Realistic concentrations for cultured cell homogenization are of the order of 5–10 mg tissue/mL rather than the 200 mg tissue/mL used for solid tissues. One device that is reasonably effective is the nitrogen bomb, which uses gas cavitation to disrupt the cells (*see* Chapter 9). The exact conditions required must be determined for each cell type. We have found that efficient (95%) disruption of Caco-2 cells (a human colonic carcinoma cell-line) can be obtained by suspending the cells in 0.25$M$ sucrose/12 m$M$ Tris-HCl, pH 7.4 and applying nitrogen at 550 psi (3.7 MPa) for 10 min. The cytoskeleton of cultured cells appears to be more highly organized than that of the corresponding tissue. This can result in the association of organelles with cytoskeletal components surrounding the nucleus, such that as much as 50% of the homogenate may sediment with the nucleus *(8)*. Confluent cells tend to develop a more complex network of cytoplasmic filaments and are even more difficult to process. It is advisable therefore to standardize and optimize the growth conditions before harvesting.

### 1.2.2. Isolation of Membrane Fraction

The choice is between isolation of a highly purified subcellular fraction enriched in the enzyme or the rapid preparation of a crude membrane fraction, free of nuclei and cytosol. If the enzyme is located in an organelle that can be purified efficiently from the tissue source used, then extensive subcellular fractionation to gain a significant degree of purification prior to solubilization may be worthwhile (*see* Chapters 2–13). The balance of yield and specific activity will determine the advisability of such a course. If, however, the location of the enzyme is not known or it is distributed over several subcellular fractions, as in the case of rat liver 5'-nucleotidase, then it is better to prepare a crude membrane fraction. A crude membrane pellet, free of cytosol and nuclei, can be obtained by homogenization in isotonic $0.25M$ sucrose buffer followed by centrifugation at $1000g$ for 10 min to sediment the nuclei and unbroken cells. The postnuclear supernatant is then centrifuged at $100,000g$ for 1 h. This should produce membranes enriched two- to fivefold relative to the homogenate, free of nuclei and cytosol. Homogenization and solubilization can result in the liberation and activation of proteases, and it is therefore often advisable to include protease inhibitors at these steps. The metal chelators EDTA and EGTA used at 1–5 m$M$ inhibit metalloproteases and also guard against heavy metal poisoning. Serine proteases are inhibited by 1 m$M$ phenylmethylsulfonyl fluoride or benzamidine. Thiol proteases are inhibited by 1 m$M$ iodoacetamide or $N$-ethylmaleimide. Aspartyl proteases are inhibited by 10 $\mu M$ pepstatin A. The inclusion of 1 m$M$ dithiothreitol may also be necessary to prevent oxidation of essential thiol groups.

The pelleted membranes should be resuspended in a suitable low ionic strength buffer at a concentration of 1–5 mg/mL protein and stored in aliquots at –70°C or in liquid nitrogen for use in preliminary characterization studies of the enzyme.

### 1.2.3. Solubilization

#### 1.2.3.1. SELECTION OF SOLUBILIZATION AGENT

Integral membrane proteins require detergent solubilization, whereas peripheral proteins can be solubilized by conditions that disrupt hydrophilic interactions such as high salt concentrations (>0.15$M$), extremes of pH (pH 3–5; pH 8–12), or metal chelation.

The selection of a suitable detergent for the solubilization step is an important factor in a successful purification scheme for an integral membrane enzyme. Effective subsequent fractionation requires 1 mol of the enzyme/detergent micelle. There is a wide choice of well characterized detergents available commercially, and their properties and uses have been extensively reviewed *(10)*. The choice of detergent usually has to be made empirically since there are few predictive factors to guide the choice, although it has been suggested that proteins possessing GPI tails are more readily solubilized by detergents with high critical micelle concentrations (CMC) *(11)*. Pilot studies should test several detergents of different structural types.

Detergents are classified by their electric charge as ionic, nonionic, or zwitterionic and by their overall structure as Type A or Type B *(12)*. Type A detergents have hydrophilic head groups and flexible hydrophobic tails that form large micelles (approx 20–90 kDa) in aqueous solution at concentrations greater than the CMC. Below this concentration the detergents exist as monomers. The CMC of Type A detergents with nonionic or zwitterionic head groups is generally low (<1 m$M$), whereas ionic detergents have higher CMCs. Type B detergents (bile salts and conjugated bile salts) have rigid cholesterol-based backbones, one side of which is hydrophobic and the other hydrophilic; they form small micelles and have high CMCs. Detergents with high CMCs can be removed from solution by dialysis, removal of detergents with low CMCs requires detergent-binding beads such as Bio-Rad SM2 beads *(13)*. A useful initial selection of detergents might include:

*Type A detergents*
Nonionic:      octyl glucoside; Triton X-100; Lubrol PX
Zwitterionic:  sulfobetaine 14 (Zwittergent 3-14, *N*-tetradecyl-*N*,*N*-dimethyl-3-ammonio-1-propanesulfonate)

*Type B detergents*
Nonionic: digitonin
Zwitterionic:  CHAPS (3-[cholamidopropyl]dimethyl-3-ammonio-1-propanesulfonate)
Ionic:      sodium cholate

### 1.2.3.2. Pilot Solubilization Study

Suspend membranes at 1–5 mg/mL protein. Add detergent to a final concentration of 0.1–5% (w/v), with a 20:1 ratio of detergent to membrane protein (w/w) at the highest concentration used. Incubate 30–60 min at 4°C. Centrifuge at 100,000g for 1 h. Determine the enzyme activity and protein concentration in an uncentrifuged aliquot, and in the supernatant, and in the resuspended pellet from the centrifugation. The addition of 0.15–0.5M NaCl to the detergent may aid solubilization since ionic interactions may also be involved even with integral membrane proteins. Mere sedimentation at these rcfs is not a sufficient criterion of solubilization, and some analysis of the size of the dispersed protein is advisable *(14)*. This can be provided by gel filtration or nondenaturing polyacrylamide gel electrophoresis *(15)*.

Once a suitable detergent has been chosen, it is useful at this stage to determine the pH stability and salt tolerance of the solubilized enzyme, since this information will be of use in designing the subsequent purification steps. In scaling up the solubilization step for full-scale purification, the concentration of detergent required may need to be increased since the ratio of detergent to protein is also an important parameter.

### 1.2.3.3. Use of Triton X-114

An alternative approach is to use the phase transition properties of the nonionic detergent Triton X-114 to distinguish between integral and peripheral proteins. This detergent undergoes a phase transition at temperatures >20°C into a detergent-rich phase and an aqueous phase. In general, integral membrane proteins partition into the detergent phase, whereas soluble and peripheral proteins partition into the aqueous phase *(9)*. However, examples of anomalous sorting do occur, and this method should not be the sole criterion used. A protocol for Triton X-114 extraction is as follows:

1. Incubate the membranes (1–5 mg/mL protein) with 1–2% of the detergent for 30–60 min at 4°C, and pellet the insoluble material by centrifugation at 100,000g for 1 h.
2. Incubate the supernatant for 3–5 min at 30°C, the solution should turn cloudy, indicating that phase transition has occurred. If it does not, try adding additional detergent.

3. Centrifuge for 3–5 min at >20°C, >1000g to separate the detergent and aqueous phases. The detergent phase forms at the bottom of the tube.

### 1.2.4. Initial Purification Steps

Before proceeding to an affinity step, it is usually advisable to increase the specific activity of the sample by one or more separation steps based on general protein features such as size, charge, or type of glycosylation since nonspecific protein binding to the "inert" matrix of the affinity step may otherwise decrease significantly the potential scope for purification. Maintenance of the enzyme in solution after solubilization requires the continued presence of detergent in the buffers subsequently used. The detergent should be used at a concentration greater than the CMC; for uncharged detergents, 0.05–0.1% detergent is usually sufficient.

### 1.2.4.1. Size Separation

Gel filtration has the advantage that an estimate of the size of the native enzyme/detergent complex can be easily obtained by calibrating the column with mol-wt marker proteins but it requires a small sample volume. The resolving power of gel filtration is less at low ionic strength, and the inclusion of, e.g., $0.1M$ NaCl in the column buffer is advisable. Increased ionic strength should not affect the parameters of uncharged detergents but will decrease the CMC of charged detergents. Recovery of activity at this step should be essentially 100%, and any significant loss of activity requires investigation since likely causes will operate on other purification steps. Loss of activity could be owing to the removal of a low $M_r$ cofactor essential for activity or to oxidation of essential SH groups. The latter may be prevented or reversed by the presence of 1 m$M$ dithiothreitol.

### 1.2.4.2. Charge Separation

For ion-exchange chromatography, an uncharged detergent is preferred to ionic detergents. Depending on what is known with regard to the isoelectric point of the enzyme, a first step would be application of the enzyme at low ionic strength at a pH in the range 5–9 to a DEAE-column equilibrated in the same buffer. If the enzyme binds, it should elute with a gradient of increasing ionic strength. Alternatively, if the enzyme does not bind and a high proportion of the accompanying protein does, significant purification can also be obtained.

### 1.2.4.3. LECTIN CHROMATOGRAPHY

Lectins are proteins that bind reversibly to specific sugar residues, and immobilized lectins have been used extensively in the purification of glycoproteins *(16)*. The bound glycoprotein is eluted by a competing sugar, and good recovery of enzyme is usually possible. Immobilized lectins are available commercially and among the most commonly used in membrane protein purification are concanavalin A and lentil lectin, both of which bind α-D-mannopyranosyl and α-D-glycopyranosyl plus sterically related residues, and wheat germ lectin, which binds *N*-acetyl-D-glucosaminyl residues. In general, the sample is applied to the resin packed in a small column at a low flow rate to allow time for binding to occur. The buffer should contain NaCl (0.5$M$) to limit ion-exchange behavior. The binding of concanavalin A and lentil lectin requires bound $Ca^{2+}$ and $Mn^{2+}$, and thus metal chelators cannot be used. The column should be washed for about 10 column vol to remove non-specifically bound protein. Elution of bound protein is achieved by a gradient of competing sugar (0–0.5$M$ of α-D-methylmannoside or α-D-methylglucoside in the case of concanavalin A and lentil lectin; *N*-acetyl-glucosamine for wheat germ lectin). Once the sugar concentration at which elution is obtained is found, then step elution may be performed.

### *1.2.5. Affinity Purification*

### 1.2.5.1. CHOICE OF MATERIAL

Affinity chromatography is a very powerful technique based on specific reversible interactions between biomolecules. The most specific of these is likely to be an antibody/antigen interaction, and if a specific monoclonal antibody is available it provides probably the best route to the preparation of pure enzyme. The raising of a monoclonal antibody can be achieved using quite low purity material, provided a suitable screening assay can be devised (*see* Chapter 5 of *Biomembrane Protocols: II. Architecture and Function*). In this laboratory three monoclonal antibodies to rat liver 5'-nucleotidase were raised using material only approx 10% pure *(17,18)*, and antibodies to isoenzymes of human alkaline phosphatases were obtained using material only approx 1% pure *(19,20)*. Successful immunoaffinity purification of 5'-nucleotidase has also been achieved using specific polyclonal antibodies *(6)*, but in general, monoclonal antibodies are the reagents of choice.

The purified monoclonal antibody is immobilized on a solid support, usually CNBr-activated Sepharose 4B (Pharmacia) or Affigel (Bio-Rad), giving 1–10 mg protein/mL gel. Alternatively, the antibody can be coupled to derivitized cellulose, which gives a high capacity immunoadsorbent (0.3 mg protein/mg cellulose), though without significant flow properties. It is therefore used in a batch protocol. The rate of application of the enzyme to the immobilized antibody should allow sufficient time for binding to go to completion, and this can vary with the affinity of the antibody.

The alternative approach to an immunoaffinity step is to use an immobilized affinity ligand related to a substrate or inhibitor of the enzyme or to an enzyme cofactor. Many of these are group-specific ligands, which are commercially available. Examples include immobilized NAD/NADP for dehydrogenases and immobilized amino acids for proteases. Elution is generally with competing ligand or a related molecule. In the case of 5'-nucleotidase, affinity purification has been obtained both by (1) binding to ADP-agarose (a competitive inhibitor) and eluting with 0.5 m$M$ of the substrate AMP *(6,22)* and (2) binding to AMP-Sepharose and eluting with 10 m$M$ AMP *(23)*.

### 1.2.5.2. Elution Conditions

The most frequently used elution conditions are extremes of pH and chaotropic ions/high-ionic-strength. The pH stability profile of the enzyme should guide whether the enzyme activity will best survive high or low pH. Many membrane proteins have been eluted in an active state by alkaline pH; successful conditions include 0.05$M$ diethylamine, pH 11.2–11.5; 0.6$M$ 2-amino-2-methyl-1-propanol, pH 10.2; 0.2$M$ glycine-OH$^-$, pH 10, 0.5$M$ NaCl; 0.02$M$ lysine-OH$^-$, pH 11; 0.1$M$ triethylamine, pH 10–11.5. Low pH conditions include 0.05–0.1$M$ glycine-HCl, pH 2.5–2.8; 0.2$M$ acetic acid, pH 2.5–3; 9$M$ ethanediol, 0.3$M$ NaCl, 0.1$M$ citric acid, pH 2. To maximize recovery of active enzyme, rapid neutralization of the eluted protein may be necessary. This can be achieved by collecting the samples directly into tubes containing an appropriate buffer. Chaotropic ions/high-ionic-strength conditions include 1–3$M$ NaCl; 2–3$M$ MgCl$_2$; 4–8$M$ urea; 6$M$ guanidine-HCl; 1$M$ KCl, 0.05$M$ Na-acetate, pH 5.5. A full reference list of individual proteins eluted by the above conditions will be found elsewhere *(21)*. The eluate should also include the detergent, and therefore the possible effect of the conditions on the detergent should be borne in mind. Carboxylic acid-containing detergents such as *N*-lauryl sarcosinate and the bile acids protonate and precipitate at acidic pH. The zwitterionic bile salt derivative CHAPS is an exception owing to the low pKa of its acid group. The carboxylic acid-containing detergents also precipitate with divalent cations. However, bile salt derivitives such as CHAPS and the taurine derivatives of cholic acid and deoxycholic acid do not precipitate with divalent cations.

## 2. Materials

### *2.1. Homogenization, Isolation of Membrane Pellet, and Solubilization of Enzyme*

1. Rat liver (150–200 g).
2. Buffer A: 100 m$M$ Tris-HCl, pH 7.5, 1 m$M$ MgCl$_2$, 1 m$M$ AMP.
3. 20% (w/v) Triton X-100.
4. Waring blender.
5. Cheesecloth.
6. Ultracentrifuge with large capacity fixed-angle (e.g., 10 × 100 mL with appropriate tubes).
7. Motor driven Potter-Elvehjem homogenizer (large volume).
8. 25% (w/v) sulfobetaine (Zwittergent 3-14, Calbiochem).
9. Magnetic stirrer.
10. High-speed centrifuge with 8 × 50 mL fixed-angle rotor and appropriate tubes.

### *2.2. Ion-Exchange Chromatography*

1. NaCl.
2. Triton X-100.
3. DEAE-cellulose column (Whatman DE-52 microgranular ion-exchanger), bed vol approx 100 mL.
4. Buffer B: 100 m$M$ Tris, 85 m$M$ HCl, 100 m$M$ NaCl, 1 m$M$ MgCl$_2$, 0.05% (w/v) sulfobetaine 14, 0.05% (w/v) Triton X-100.

### *2.3. Immunoaffinity Chromatography*

1. Anti-5'-nucleotidase antibody, 5N-E5, (*see* refs. *3,5,17,18* for more information about this antibody) coupled to CNBr-activated Sepharose 4B (Pharmacia). Use the procedure recommended by manufacturers for coupling, and prepare a column vol of 16 × 0.9 cm.
2. Buffer B: *see* Section 2.2.
3. 50 m$M$ diethylamine.

4. Buffer C: 50 m$M$ diethylamine, 0.05% (w/v) sulfobetaine, 0.05% (w/v) Triton X-100.
5. Buffer D: 2$M$ Tris, 1.95$M$ HCl, 10 m$M$ MgCl$_2$.
6. Buffer E: 100 m$M$ Tris, 85 m$M$ HCl, 1 m$M$ MgCl$_2$, 0.05% (w/v) sulfobetaine 14, 0.05% (w/v) Triton X-100.
7. Anti-5'-nucleotidase monoclonal antibody (5N-E5) coupled to diazo-cellulose (*see* Chapter 13).
8. Refrigerated low-speed centrifuge with swing-out rotor for 10- to 15-mL tubes.

## 3. Methods (*see* Note 1)

Unless otherwise stated perform all steps at 4°C. All g-forces are g$_{av}$.

### 3.1. Homogenization and Isolation of Membrane Pellet and Solubilization of Enzyme (see Note 2)

1. Homogenize 150–200 g rat liver in approx 300 mL Buffer A, using a Waring blender (30–60 s).
2. Add a further 300 mL Buffer A; rehomogenize for 30 s and adjust volume to 900 mL. Add 20 mL 20% (w/v) Triton X-100 slowly with stirring.
3. Filter the homogenate through cheesecloth to remove unbroken cells and connective tissue. Then centrifuge for 90 min at 50,000$g$ in a fixed-angle rotor.
4. Aspirate and discard the supernatant, retaining the pellet and the associated loose fluffy layer. Resuspend it in a minimum volume of Buffer A (approx 200 mL) using a motor driven Potter-Elvehjem homogenizer.
5. Add 25% (w/v) sulfobetaine 14 to a final concentration of 2%, and stir for 30 min at room temperature. Centrifuge for 30 min at 35,000$g$, and collect the supernatant containing the solubilized enzyme (*see* Note 3).

### 3.2. Ion-Exchange Chromatography (see Note 4)

1. Equilibrate a DEAE-cellulose column (bed vol approx 100 mL) in Buffer B at a flow-rate of 0.5–1.0 mL/min.
2. Add NaCl and Triton X-100 to the solubilized enzyme to final concentrations of 100 m$M$ and 2% (w/v) respectively, and apply to the column.
3. Collect the unbound material, which contains the enzyme.

### 3.3. Immunoaffinity Chromatography (see Notes 5 and 6)

1. Apply the unbound material from Section 3.2. to a column of 5N-E5-Sepharose 4B, pre-eluted with 5 mL 50 m$M$ diethylamine and equilibrated in 50 mL Buffer B, at a flow-rate of 0.5–1.0 mL/min.

Table 1
Summary of Immunoaffinity Purification of Rat Liver 5'-Nucleotidase

| Step | U/mg | U | Yield (%) |
|------|------|---|-----------|
| Homogenization | 0.025 | 485 | 100 |
| Pellet | N.D. | 412 | 85 |
| Solubilization | N.D. | 354 | 73 |
| DE52/5N-E5 sepharose | 245 | 265 | 55 |
| 5N-E5 cellulose | 380 | 232 | 48 |

2. Wash the column with several column vol of Buffer B (at least 50 mL).
3. Elute the bound enzyme with ice-cold Buffer C at a flow rate of 1 mL/min, and collect 0.9-mL fractions in tubes containing 0.1 mL of Buffer D for immediate neutralization.
4. Assay the eluted fractions for the enzyme (*see* Chapter 1), and pool those containing enzyme activity; then dialyze against Buffer E.

Final purification of 5'-nucleotidase may require repeating the immunoaffinity step using the antibody coupled to diazocellulose.

5. Pre-elute 15 mg of this immunoadsorbent with cold Buffer C and equilibrate with Buffer B batchwise, using centrifugation at 1000g for 2 min to pellet the cellulose.
6. Incubate the immunoadsorbent pellet with enzyme for 60 min; then centrifuge at 1000g for 2 min.
7. Wash three times with Buffer B (5–10 mL) and incubate for 5 min with 5 mL of ice-cold Buffer C.
8. Centrifuge at 1000g for 10 min; remove the supernatant and neutralize it immediately with 0.5 mL of Buffer D.
9. Resuspend the cellulose pellet in a further 5 mL of Buffer C, and repeat immediately the centrifugation and neutralization. Then assay the two eluted fractions and pool if required.
10. Dialyze the enzyme against Buffer E.

## 4. Notes

1. This protocol results in a 15,000-fold increase in the specific activity (U enzyme activity/mg protein) of rat liver 5'-nucleotidase with an overall yield of 48% relative to the homogenate (Table 1). Protein was measured by the method of Lowry *(24)* with bovine serum albumin as standard and 1 U enzyme activity = 1 μmol AMP hydrolyzed/min at 37°C. However, 50 μg of purified enzyme by the Lowry method was

only equivalent to 21.5 µg by amino-acid compositional analysis; the true specific activity is thus of the order of 880 U/mg enzyme *(3)*.

2. The homogenization step was based on the method of Widnell et al. *(25)*, which we followed initially to produce partially purified enzyme for immunization. It is often helpful to examine the literature for related purification protocols in planning a new purification. Triton X-100 was added at low concentration to release soluble proteins from organelles, AMP was added to provide substrate protection, and $Mg^{2+}$ is required for enzyme activity.

3. Once pilot studies have established that the chosen detergent gives a truly soluble enzyme, it is often not necessary to spin at $100,000g$ for 1 h in preparative work, and when large volumes of materials are processed such a g force is often not easily attainable. Room temperature was chosen to speed solubilization, as the enzyme is very resistant to inactivation.

4. The ion-exchange step was adapted from an earlier purification of the enzyme *(15)*, which used only general purification methods, simply to reduce nonspecific binding to the Sepharose matrix. Mixed micelles of Triton X-100 and sulfobetaine 14 were generated at this stage because occasional precipitating-out of sulfobetaine 14 alone may occur when used for prolonged periods at 0–4°C. Since the enzyme does not bind to the DE52 column, the flow-through was generally connected directly to the antibody column to save time and effort.

5. In our original immunoaffinity purification of 5'-nucleotidase *(15)*, partial purification prior to solubilization was based on the first five steps of the method of Widnell *(25)* and resulted in the formation of phospholipid-enzyme pellet (18-fold purified over homogenate). This was then solubilized with sulfobetaine 14 and applied batchwise to 15 mg of antibody 5N-42-diazocellulose. A yield of 9–16% over the homogenate was obtained, with a specific activity of 285–340 U/mg protein *(15)*.

6. Antibody 5N-E5 *(18)* proved better for purification purposes because it was of lower affinity and gave higher yields of active enzyme. The cell-line also gave a better yield of antibody. Sepharose 4B replaced cellulose as the antibody support because suspension of the cellulose immunoadsorbent in the large volume of solubilized material generated by this protocol gave a much lower antibody concentration leading to reduced binding. Use of Sepharose packed in a column gave a high antibody concentration. The low specific activity of the material applied to the column compared to that used in the original method meant that complete purification required two immunoaffinity steps.

# References

1. Singer, S. J. (1990) The structure and insertion of integral proteins in membranes. *Ann. Rev. Cell Biol.* **6,** 247–296.
2. Cross, G. A. M. (1990) Glycolipid anchoring of plasma membrane proteins. *Ann. Rev. Cell Biol.* **6,** 1–39.
3. Bailyes, E. M., Ferguson, M. A. J., Colaco, C. A. L. S., and Luzio, J. P. (1990) Inositol is a constituent of detergent-solubilized immunoaffinity-purified rat liver 5'-nucleotidase. *Biochem. J.* **265,** 907–909.
4. Luzio, J. P., Bailyes, E. M., Baron, M., Siddle, K., Mullock, B. M., Geuze, H. J., and Stanley, K. K. (1986) The properties, structure, function, intracellular localization and movement of hepatic 5'-nucleotidase, in *Cellular Biology of Ectoenzymes* (Kreutzberg, G. W., eds.), Springer-Verlag, Berlin, Heidelberg, pp. 89–116.
5. Luzio, J. P., Baron, M. D., and Bailyes, E. M. (1987) Cell biology, in *Mammalian Ectoenzymes* (Kenny, A. J. and Turner, A. J., eds.), Elsevier/North Holland, Amsterdam, pp. 111–138.
6. Misumi, Y., Ogata, S., Hirose, S., and Ikehara, Y. (1990) Primary structure of rat liver 5'-nucleotidase deduced from the cDNA. Presence of the COOH-terminal hydrophobic domain for possible post-translational modification by glycophospholipid. *J. Biol. Chem.* **265,** 2178–2183.
7. Ogata, S., Hayashi, Y., Misumi, Y., and Ikehara, Y. (1990) Membrane-anchoring domain of rat liver 5'-nucleotidase: identification of the COOH-terminal serine-523 covalently attached with a glycolipid. *Biochemistry* **29,** 7923–2927.
8. Howell, K. E., Devaney, E., and Gruenberg, J. (1989) Subcellular fractionation of tissue culture cells. *Trends Biochem. Sci.* **14,** 44–47.
9. Pryde, J. G. (1986) Triton X-114: a detergent that has come in from the cold. *Trends Biochem. Sci.* **11,** 160–163.
10. Hjelmeland, L. M. and Chrambach, A. (1984) Solubilization of functional membrane proteins. *Methods Enzymol.* **104,** 305–318.
11. Hooper, N. M. and Turner, A. J. (1988) Ectoenzymes of the kidney microvillar membrane. Differential solubilization by detergents can predict a glycosylphosphatidylinositol membrane anchor. *Biochem. J.* **250,** 865–869.
12. Helenius, A. and Simons, K. (1975) Solubilization of membranes by detergents. *Biochim. Biophys. Acta* **415,** 29–79.
13. Metsikko, K., van Meer, G., and Simons, K. (1986) Reconstitution of the fusogenic activity of vesicular stomatitis virus. *EMBO. J.* **5,** 3429–3435.
14. Newby, A. C., Chrambach, A., and Bailyes, E. M. (1982) The choice of detergents for molecular characterisation and purification of native intrinsic membrane proteins, in *Techniques in Lipid and Membrane Biochemistry* B409 (Hesketh, T. R., eds.), Elsevier/North Holland, Amsterdam, pp. 1–22.
15. Bailyes, E. M., Newby, A. C., Siddle, K., and Luzio, J. P. (1982) Solubilization and purification of rat liver 5'-nucleotidase by use of a zwitterionic detergent and a monoclonal antibody immunoadsorbent. *Biochem. J.* **203,** 245–251.
16. Lis, H. and Sharon, N. (1986) Lectins as molecules and tools. *Ann. Rev. Biochem.* **55,** 35–67.

17. Siddle, K., Bailyes, E. M., and Luzio, J. P. (1981) A monoclonal antibody inhibiting rat liver 5'-nucleotidase. *FEBS Lett.* **128,** 103–107.
18. Bailyes, E. M., Soos, M., Jackson, P. J., Newby, A. C., Siddle, K., and Luzio, J. P. (1984) The existence and properties of two dimers of rat liver ecto-5'-nucleotidase. *Biochem. J.* **221,** 369–377.
19. Bailyes, E. M., Siddle, K., and Luzio, J. P. (1982) Monoclonal antibodies against human alkaline phosphatases. *Biochem. Soc. Trans.* **13,** 107.
20. Bailyes, E. M., Seabrook, R. N., Calvin, J., Maguire, G. A., Price, C. P., Siddle, K., and Luzio, J. P. (1987) The preparation of monoclonal antibodies to human bone and liver alkaline phosphatase and their use in immunoaffinity purification and in studying these enzymes when present in serum. *Biochem. J.* **244,** 725–733.
21. Bailyes, E. M., Richardson, P. J., and Luzio, J. P. (1987) Immunological methods applicable to membranes, in *Biological Membranes: A Practical Approach* (Findlay, J. B. C. and Evans, W. H., eds.), IRL at Oxford University Press, Oxford, pp. 73–101.
22. Naito, Y. and Lowenstein, J. M. (1981) 5'-nucleotidase from rat heart. *Biochemistry* **20,** 5188–5194.
23. Harb, J., Meflah, K., Duflos, Y., and Bernard, S. (1983) Purification and properties of bovine liver plasma membrane 5'-nucleotidase. *Eur. J. Biochem.* **137,** 131–138.
24. Lowry, O. H., Rosebrough, N. J., Farr, A. L., and Randell, R. J. (1951) Protein measurement with the Folin phenol reagent. *J. Biol. Chem.* **193,** 265–275.
25. Widnell, C. C. (1975) Purification of rat liver 5'-nucleotidase as a complex with sphingomyelin. *Methods Enzymol.* **32,** 368–374.

# CHAPTER 23

# The Isolation
# of Membrane Proteoglycans

*Malcolm Lyon*

## 1. Introduction

The proteoglycans (PGs) are a large and varied family of complex macromolecules whose physical properties are dominated by their large sulfated polysaccharide chains (for a recent review, *see* ref. *1*). They are widely distributed in the animal kingdom, but are not associated with any particular anatomical or cellular sites, being instead almost ubiquitously present both on cells and within extracellular matrices. On cells they are usually constituents of the plasma membranes, though in the case of various cells of the hemopoietic lineage they can also be found in high concentrations within secretory granules.

The composite nature of PGs, with both their core proteins and glycosaminoglycans (GAGs) varying in structure, leads to a considerable level of complexity. Approximately 15 PG core proteins have now been sequenced and more are sure to follow in the near future. In contrast, the basic categories of GAG have long been recognized (i.e., chondroitin sulfate (CS), dermatan sulfate (DS), keratan sulfate (KS), heparan sulfate (HS), and heparin) and have been used as a basis for PG classification, although the existence of hybrid PGs containing two GAG species (e.g., CS/KS or CS/HS) has blurred the distinctions. Further complications arise from the apparent lack of a template-directed polymerization mechanism akin to that of proteins, with frequently extensive modifications occurring after polymerization. The end result is that both the fine structure and chain length of

From: *Methods in Molecular Biology, Vol. 19: Biomembrane Protocols: I. Isolation and Analysis*
Edited by: J. M. Graham and J. A. Higgins Copyright ©1993 Humana Press Inc., Totowa, NJ

the GAGs may be modulated. No doubt a potential for programmable variation in structure is of considerable functional importance in directing a repertoire of putative interactions with other macromolecules. However, the corollary is that even a single PG species always behaves as a broad polydisperse population, and this leads to a consequent loss of resolution in most separation techniques. Fortunately there are a number of relatively distinct characteristics common to many PGs that can be used to some advantage for purification purposes, namely their large hydrodynamic volume, their high and effectively nontitratable polyanionic charge, and their relatively high buoyant density.

Nearly all large-scale purifications to date have concerned the extracellular matrix PGs, principally the large CSPG of cartilage and the large basement membrane HSPG. In the case of the former, its high abundance in a rather acellular tissue, together with its ability to aggregate forming massive structures, are considerable aids to purification. Similarly, with the latter HSPG, it has been possible to take advantage of relatively rich sources of basement membrane (e.g., Engelbreth-Holm-Swarm sarcoma or kidney glomeruli) where the intact basement membrane can be prepared free of much of the contaminating cellular material by washing with appropriate salt/detergent solutions.

In contrast the generally smaller cell-associated PGs are present at much lower abundance in a cellular environment containing a wide variety of potentially contaminating molecules. Cells may also express multiple species of PG. On a small scale, work with cultured cells has yielded considerable analytical information, especially when aided by the availability of specific monoclonal antibodies, but large scale purifications have been elusive. Clearly, future progress in understanding the interactive properties of these molecules and their biological function is going to depend on the availability of reagent quantities of purified PGs. In this instance, recourse to recombinant production technology is unlikely to be of value because of the requirement for the correct glycanation of the product.

This chapter describes the protocol that has been successfully employed for the first large-scale purification of a membrane PG, namely the major membrane HSPG from rat liver *(2)*. Undoubtedly, starting with other sources of material may require specific modifications to this approach, and hopefully the suggestions made in the Notes section will be of assistance in developing an appropriate strategy.

## 2. Materials

1. 0.15$M$ NaCl, 20 m$M$ sodium phosphate, pH 5.0.
2. Extraction solution: 4$M$ guanidinium chloride, 2% (v/v) Triton X-100, 50 m$M$ sodium acetate, pH 5.0 (*see* Note 2) containing 0.1$M$ 6-aminohexanoic acid, 20 m$M$ benzamidinium chloride, 10 m$M$ EDTA, and 5 m$M$ $N$-ethylmaleimide as proteinase inhibitors. Store at 4°C. High concentrations of relatively low grade guanidinium chloride will contain appreciable quantities of UV-absorbing contaminants. These are best removed by stirring an 8$M$ stock solution overnight with activated charcoal. The charcoal can be filtered out using Whatman paper, although further passage through a 0.22-µm filter will be necessary to remove the finest particles. The molarity of a guanidinium chloride stock solution can be calculated from the density using Eq. (1).

$$\text{Molarity} = (\rho - 1.003)/0.02359 \tag{1}$$

where $\rho$ is density (g/cm$^3$).
3. 0.2$M$ Phenylmethylsulfonylfluoride in methanol (PMSF). **Caution:** highly toxic!
4. Dounce homogenizer or electric blender (*see* Note 1).
5. 100% (w/v) TCA. Store at 4°C.
6. A high capacity low-speed refrigerated centrifuge capable of handling the total volume of TCA supernatant, preferably in a single run.
7. 10$M$ NaOH solution. **Caution:** highly corrosive!
8. DEAE-Sephacel.
9. MonoQ HR5/5 column linked to a Pharmacia FPLC or similar system. A small DEAE-Sephacel column run conventionally may suffice, though resolution is likely to be inferior.
10. Solid poly(ethylene glycol) or carboxymethylcellulose.
11. Solutions for DEAE-Sephacel chromatography.

Solution A:   8$M$ urea, 0.15$M$ NaCl, 0.5% (v/v) Triton X-100, 20 m$M$ Tris-HCl, pH 8.0.

Solution B:   6$M$ urea, 0.15$M$ NaCl, 0.5% (v/v) Triton X-100, 20 m$M$ piperazine-HCl, pH 5.0.

Solution C:   6$M$ urea, 1$M$ NaCl, 0.5% (v/v) Triton X-100, 20 m$M$ piperazine-HCl, pH 5.0.

Use freshly prepared 8$M$ urea stock solutions, deionized by stirring overnight with a mixed anion/cation exchange resin (e.g., Amberlite MB-3), and remove the resin by filtration. Store the stock solution at 4°C. Once salts/buffers have been added to make the working solutions they should be used within a week of preparation.

12. Chondroitinase ABC buffer: 50 m$M$ NaCl, 50 m$M$ Tris-HCl, 0.1% (v/v) Triton X-100, pH 8.0 containing 10 m$M$ 6-aminohexanoic acid, 10 m$M$ benzamidinium chloride, 10 m$M$ EDTA, and 5 m$M$ $N$-ethylmaleimide as proteinase inhibitors.

13. Chondroitinase ABC (chondroitin ABC lyase from *Proteus vulgaris*; E.C. 4.2.2.4). Store at –20°C.

14. Solutions for MonoQ chromatography (*see* item 11 above for preparation).
    Solution A:  6$M$ urea, 0.15$M$ NaCl, 0.1% (v/v) Triton X-100, 20 m$M$ Tris-HCl, pH 8.0.
    Solution B:  6$M$ urea, 0.15$M$ NaCl, 0.5% (w/v) CHAPS, 20 m$M$ Tris-HCl, pH 8.0.
    Solution C:  6$M$ urea, 1.5$M$ NaCl, 0.5% (w/v) CHAPS, 20 m$M$ Tris-HCl, pH 8.0.

15. 4$M$ Guanidinium chloride, 0.2% (w/v) CHAPS, 20 m$M$ sodium phosphate, pH 7.3.

16. CsCl of high quality, e.g., ultracentrifuge or molecular biology grade.

17. Ultracentrifuge preferably equipped with a titanium rotor. We use a fixed-angle rotor, though vertical rotors could be used. Heat-sealable centrifuge tubes are a considerable advantage.

18. Centrifuge tube emptying device. We prefer to empty from the bottom by piercing with a needle and withdrawing the contents using a peristaltic pump. It is important that the total length of tubing is minimized, and that a continuous downward direction of flow is maintained to reduce gradient mixing during tube emptying.

19. Absolute ethanol containing 1.3% (w/v) potassium acetate.

20. Gradient making device.

21. Millipore™ filters (0.22 µm).

## 3. Methods

All g-forces are g$_{av}$.

### 3.1. Extraction

The procedure described was developed for the handling of 30 whole rat livers, three of which had been metabolically radiolabeled by an intraperitoneal injection of the animal with Na$_2$$^{35}$SO$_4$ 2 h before sacrifice. The radiolabel is followed as a proteoglycan marker. If other tissues, plasma membrane preparations, or cultured cells are to be used as starting material then substantial modifications may be necessary (*see* Notes 1 and 2).

1. Rinse excised livers in ice-cold 0.15$M$ NaCl, 20 m$M$ sodium phosphate, pH 5.0 to remove as much blood as possible.

2. Homogenize the livers in ice-cold extraction solution (50–80 mL/liver) (supplemented with 0.5 m$M$ PMSF immediately before use) using a loose fitting Dounce homogenizer or an electric blender.
3. Stir the homogenate at 4°C for 24 h. Remove any large insoluble residues by filtration through muslin or nylon mesh.

### 3.2. TCA Precipitation

The solubility of the HSPG in TCA/guanidinium chloride allows for easy removal of the majority of nucleic acid and protein (*see* Note 3).

1. Add sufficient 100% (w/v) TCA to the ice-cold extract to give 10% (w/v) final concentration. Mix well during the addition. After 15 min at 0°C, centrifuge at 2000$g$ and 4°C for 30 min. Remove the supernatant and neutralize by careful addition, with mixing, of 10$M$ NaOH solution.
2. Wash the TCA precipitate with fresh extraction solution containing 10% (w/v) TCA and recentrifuge. Recover the wash supernatant, neutralize as above, and pool with the original supernatant.
3. Concentrate the pooled supernatant by reverse osmosis in dialysis bags against solid poly(ethylene glycol) or carboxymethyl cellulose (*see* Note 4) Dialyze the concentrate extensively against the 8$M$ urea solution (DEAE-Sephacel solution A).

### 3.3. Ion-Exchange Chromatography

This is performed in two stages with an intervening enzymic degradation of unwanted CS/DSPGs (*see* Notes 5–9).

1. Equilibrate approx 200 mL of DEAE-Sephacel in solution A. Add the dialyzed TCA supernatant and gently mix the slurry at 4°C overnight (*see* Note 5).
2. Centrifuge the slurry at 300$g$ for 10 min. Remove the supernatant. Wash the gel with fresh solution A and recentrifuge. Remove wash supernatant. Form a gel slurry with fresh solution A and pack into a column. Wash column with 2 vol of solution A.
3. Switch to a low pH wash using 2 or 3 vol of solution B (*see* Note 6).
4. Elute PGs with a 0.15–1.0$M$ NaCl gradient formed from solutions B and C respectively. Collect fractions and monitor $A_{280}$ and $^{35}$S content. Liver HSPG elutes at approx 0.5$M$ NaCl.
5. Pool appropriate fractions and concentrate to <10 mL against poly(ethylene glycol) or carboxymethylcellulose.
6. Dialyze concentrate against chondroitinase ABC buffer. Add 0.1 U of chondroitinase ABC, together with 0.25 m$M$ PMSF, and incubate at 37°C for at least 2 h.

7. Clarify digest by centrifugation at 3000$g$ for 10 min followed by passage through a 0.22-μm filter. Load onto the MonoQ column equilibrated in solution A and wash with this solution. Exchange detergents from Triton X-100 to CHAPS by continuing to wash with several volumes of solution B (*see* Note 9).

8. Elute PGs on a 0.15–1.5$M$ NaCl gradient formed from solutions B and C respectively. Collect fractions and monitor A$_{280}$ and $^{35}$S content. Liver elutes at about 1$M$ NaCl. Pool appropriate fractions.

### *3.4. Isopycnic Density-Gradient Centrifugation*

The relatively high buoyant density of the HSPG can now be utilized to allow separation from residual protein (*see* Notes 10–12).

1. Dialyze sample against 4$M$ guanidinium chloride, 0.2% (w/v) CHAPS, 20 m$M$ sodium phosphate, pH 7.3.

2. Adjust to a density of 1.5 g/cm$^3$ and a volume of 12 mL by the addition of solid CsCl and 4$M$ guanidinium chloride solution respectively (*see* Note 10). Centrifuge at 3000$g$ for 10 min to remove any turbidity.

3. Centrifuge at 130,000$g$ for approx 3 d at 15°C.

4. Carefully empty the tube from the bottom using a needle (*see* Materials) and collect fractions. Monitor these for density, A$_{280}$ and $^{35}$S content. Liver HSPG bands at a density of approx 1.6 g/cm$^3$. Pool appropriate fractions.

5. Precipitate purified HSPG by addition of 4 vol of absolute ethanol, 1.3% (w/v) potassium acetate at –20°C for 4 h. Selectively resolubilize the CsCl from the large pellet by repeated cycles of washing with 75% ethanol and recentrifugation (*see* Note 11). Air-dry the remaining HSPG pellet and dissolve in an appropriate buffer solution for storage at –20°C.

### 4. Notes

1. Moderate quantities of soft tissue can be adequately dealt with using a Dounce homogenizer, although for larger quantities an electric blender can be useful. Fibrous tissues may need cutting into small pieces for extraction, or freezing in liquid N$_2$ and pulverizing to a powder. In contrast, cultured cells or plasma membrane preparations may be readily solubilized without special treatment.

2. A pH of 5.8 is commonly used during guanidinium chloride extractions to minimize the activities of both neutral and acid proteases. In the present case, pH 5.0 was adopted in order to inhibit specifically a potent HS-degrading endoglycosidase associated with liver membranes *(3)*.

3. TCA precipitation has been an invaluable step in facilitating this purification. Without it the very high nucleic acid content of the liver extract

makes application to an ion-exchange gel impossible. Clearly, in cases (e.g., plasma membranes) where the nucleic acid content is relatively low this step is unnecessary. However, if the use of TCA is contemplated it must be borne in mind that the solubility in TCA/guanidinium chloride of other proteoglycans is unknown. This may depend on the protein/GAG ratio of individual PGs. It is therefore essential that the behavior of a particular PG is ascertained before proceeding on a large scale.

4. Volume reduction at all stages could be effected using ultrafiltration cells, if available, although membrane blockage may be a problem with crude samples in the early stages of purification.

5. Batch adsorption to DEAE-Sephacel was adopted because of difficulties in applying the sample to a packed column. However in circumstances where the latter can be achieved the subsequent resolution is likely to be improved. The volume of gel used is clearly dependent on the scale of purification.

6. A pH shift from 8.0 to 5.0 during washing of the DEAE-Sephacel results in significant elution of weakly bound material without affecting the PG affinity.

7. Ion-exchange does not resolve liver HSPG from CS/DSPGs. The most convenient method of removal of CS/DSPG is therefore by specific degradation using chondroitinase ABC, followed by removal of the degradation products using a second ion-exchange chromatography step. In other situations there may be sufficient resolution of PG types by ion-exchange chromatography alone. Specific degradation is nevertheless a convenient way of simplifying a complex separation problem where one species or type of PG is to be isolated and any others can be discarded. Specific reagents to remove HS (e.g., heparinases or nitrous acid) or KS (e.g., keratanase) could also be utilized.

8. An alternative to a second ion-exchange step may be gel filtration in 4$M$ guanidinium chloride and detergent. If the particular PG is very large this can prove of great benefit. This can also be a good method for removing large nucleic acid molecules. However, the inherent polydispersity of PG populations may lead to relatively poor separation between PG species and only small improvements in overall protein purity, coupled with recovery in a large volume. In the case of the liver HSPG, the MonoQ step gave a much more rapid and better purification than gel filtration on Sepharose CL6B, as well as leading to recovery of PG in a small volume.

9. MonoQ can be conveniently used to effect a change in detergent from Triton X-100 to CHAPS. The latter has a much higher critical micelle concentration and a considerably smaller micelle size. This reduces the

amount of detergent binding to the PG thereby increasing its buoyant density and maximizing the density differential between the PG and the mostly low density protein contaminants. CHAPS could theoretically be used throughout the purification, but it is considerably more expensive than Triton X-100 for large-scale use.

10. The buoyant density of PGs in CsCl/4$M$ guanidinium chloride depends primarily on the GAG/protein ratio and the GAG type. The behavior of individual PG species and the optimum conditions for resolution can only be determined by experimentation. Separation of multiple species of PGs may require more than one run under different gradient conditions. The weight of CsCl to be added to a sample in 4$M$ guanidinium chloride to achieve a target density and volume can be calculated using Eq. (2).

$$CsCl \ (gram) = V \ (1.347\rho - 1.474) \qquad (2)$$

where $\rho$ is density (g/cm$^3$) and V is volume (mL). The target volume is then obtained by addition of 4$M$ guanidinium chloride solution until the expected weight (V$\rho$) is achieved. It is highly desirable to consult a centrifugation handbook (e.g., ref. *4*) before embarking on experimentation unless an established protocol is being exactly reproduced.

11. PG can also be recovered by dialysis into an appropriate buffer. However, ethanol precipitation gives more control over the final stock concentration. Initial coprecipitation of CsCl can be avoided by appropriate dilution of the sample with water, if desired.

12. Analysis by SDS-PAGE (*see* Chapter 19) will be necessary at each stage to assess the degree of purity attained. Using the protocol described here we obtained approx 1.5 mg of apparently homogeneous, high-mol-wt liver HSPG (Fig. 1A) with a recovery of at least 25%. Enzymic deglycosylation of the HSPG yielded three related core proteins with $M_r$ of 77,49, and 44 kDa (Fig. 1B). In addition, assessment of hydrophobicity may be useful to differentiate between integral and peripheral membrane PGs. This can be undertaken either by analysis of PG binding to Octyl-Sepharose under nondenaturing conditions *(3)*, or by electrophoretic detergent blotting *(5)*, which we have found to be particularly useful where a complex mixture of core proteins is present *(2)*. Although not used as part of our liver protocol, hydrophobic interaction chromatography may be very useful as an additional purification method for membrane PGs (for further details, *see* ref. *6*).

## Acknowledgment

This work was supported by the Cancer Research Campaign.

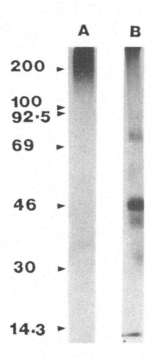

Fig. 1. SDS-PAGE analysis of $^{125}$I-labeled purified rat liver HSPG. Samples were run on a 7.5% acrylamide gel under reducing conditions. Lane **A**: intact HSPG. Lane **B**: core proteins released by digestion with heparitinase (a residual amount of undegraded PG remains at the top of the gel). The migration positions of $^{14}$C-protein mol-wt markers are shown.

## References

1. Gallagher, J. T. (1989) The extended family of proteoglycans: social residents of the pericellular domain. *Curr. Opinion Cell Biol.* **1,** 1201–1218.
2. Lyon, M. and Gallagher, J. T. (1991) Purification and partial characterisation of the major cell-associated heparan sulphate proteoglycan of rat liver. *Biochem. J.* **273,** 415–422.
3. Gallagher, J. T., Walker, A., Lyon, M., and Evans, W. H. (1988) Heparan sulphate-degrading endoglycosidase in liver plasma membranes. *Biochem. J.* **250,** 719–726.
4. Rickwood, D. (ed.) (1984) *Centrifugation: A Practical Approach* (2nd ed.), IRL at Oxford University Press, Oxford.
5. Ito, K. and Akijama, Y. (1985) Protein blotting through a detergent layer, a simple method for detecting integral membrane proteins separated by SDS-polyacrylamide gel electrophoresis. *Biochem. Biophys. Res. Commun.* **133,** 214–221.
6. Yanagishita, M., Midura, R. J., and Hascall, V. C. (1987) Proteoglycans: isolation and purification from tissue cultures, in *Methods in Enzymology*, vol. 138 (Ginsburg, V., ed.), Academic, New York, pp. 279–289.

CHAPTER 24

# Oligosaccharide Mapping and Sequence Analysis of Glycosaminoglycans

## *Jeremy E. Turnbull*

## 1. Introduction
### *1.1. General Strategy*

A strategy for determining the structure of a glycosaminoglycan (GAG) can be divided into three areas. The first is to establish its saccharide composition. This requires complete depolymerization of the parent chains to disaccharide products that can be purified, separated, and identified either by comparison with reference standards or by further structural analysis *(1–6)*.

Second, information on disaccharide distribution can be obtained by characterizing their arrangement into domains with defined structural features. Depolymerization of GAG chains with a single linkage-specific reagent generates oligosaccharides that contain end-groups that constituted the susceptible linkages, and internal sequences defined by their resistance to the reagent. Separation of these products allows assessment of the distribution of susceptible linkages in the intact chain, a type of analysis called oligosaccharide mapping *(5,7,8)*. This provides a molecular fingerprint of the polymeric structure and yields detailed structural information *(see* Section 1.2.1.). Oligosaccharide mapping can also provide oligosaccharide products that are amenable to further analysis *(see* Section 1.2.2.).

A number of methods are available for the cleavage of GAG chains at specific linkages. In general the enzymic methods, principally the polysaccharide lyases *(see* Notes 1 and 2), have the advantage of more

From: *Methods in Molecular Biology, Vol. 19: Biomembrane Protocols: I. Isolation and Analysis*
Edited by: J. M. Graham and J. A. Higgins Copyright ©1993 Humana Press Inc., Totowa, NJ

Table 1
Specificities of the Polysaccharide Lyases

| Lyase | Linkage specificity[a] |
|---|---|
| Heparitinases[b] | |
| Heparitinase I | GlcNR(±6S) α1-4GlcA |
| Heparitinase II | GlcNR(±6S) α1-4GlcA/IdoA |
| Heparitinase III | GlcNSO₃(±6S) α1-4IdoA(2S) |
| | |
| Heparinase | GlcNSO₃(±6S) α1-4IdoA(2S) |
| Heparinase II | GlcNR(±6S) α1-4GlcA/IdoA |
| Heparinase III | GlcNR(±6S) α1-4GlcA |
| | |
| Chondroitinases | |
| Chondroitinase ABC | GalNAc(±4S,6S) β1-4GlcA/IdoA(±2S) |
| Chondroitinase ACI | GalNAc(±4S,6S) β1-4GlcA |
| Chondroitinase ACII | GalNAc(±4S,6S) β1-4GlcA |
| Chondroitinase B | GalNAc(4S) β1-4IdoA(±2S) |

[a]GlcNAc, *N*-acetylglucosamine; GlcNSO₃, *N*-sulfated glucosamine; GlcNR (R = Ac or SO₃); GlcA, glucuronic acid; IdoA, iduronic acid; GalNAc, *N*-acetylgalactosamine.

[b]These enzymes are called either heparitinases or heparinases, depending on the commercial supplier: The commonly used heparitinase (heparitinase I) supplied by Seikagaku is designated heparinase III by Sigma. Similarly, heparitinase II is known as heparinase II, and heparitinase III as heparinase. These three pairs of enzymes appear to exhibit identical substrate specificities as shown in the table (*see* ref. *23* for further details). For details of resistant sequences, *see* ref. *5*.

restricted linkage specificities (*see* Table 1) than the chemical methods (*see* Notes 3 and 4), and can thus give more detailed structural information. In addition, they result in products that retain the sulfation patterns of the original sequences and are therefore well suited to further structural analysis (*see* Section 1.2.2.).

Finally, a number of approaches can be used to investigate the actual sequence of sugars from defined reference points (*see* Sections 1.2.2. and 1.2.3.). This chapter will focus principally on methods for mapping domain structure (*see* Sections 1.3. and 1.4.).

## *1.2. Analytical Strategy*

### *1.2.1. Analysis of Mapping Data*

Gel filtration profiles can be used quantitatively to calculate the proportion of linkages susceptible to a particular scission technique (*see* ref. *5* for details), and to give information on their distribution within the intact GAG chain, i.e., contiguous, alternating, or spaced apart by nonsusceptible sequences *(5)*. The latter oligosaccharides can also provide useful information on the arrangement of particular types of resistant sequences, e.g., heparitinase I resistant oligosaccharides in heparan sulfate represent contiguous sequences of iduronic acid (IdoA)-containing disaccharides *(5,8); see also* Fig. 1.

Oligosaccharide mapping by both gradient polyacrylamide gel electrophoresis (PAGE) and high-performance liquid chromatography (HPLC) are particularly useful for revealing the structural complexity of oligosaccharide mixtures, and they also provide a rapid and reproducible means of making detailed comparisons between different species of a particular GAG type. Gradient PAGE mapping is the most powerful technique available for resolving GAG oligosaccharides, and although it should not be used as a quantitative method, it can be applied usefully to the detailed resolution of oligosaccharides, which have been quantitated and broadly sized by an initial gel filtration step (e.g., *see* Fig. 1).

### 1.2.2. Secondary Structural Analysis

Following primary cleavage of the chains, further analyses of resistant sequences, prepared by gel filtration, anion-exchange HPLC, or preparative gradient PAGE methods can be made. Secondary reagents that cleave at different linkages can be used. The distribution of these linkages within the resistant oligosaccharides of defined size can be established, providing more detailed information on the composition and sequence of specific domains. For example, the distribution of IdoA(2S) residues within IdoA-repeat (heparitinase-resistant) sequences in heparan sulfate has been investigated in this manner *(8)*.

### 1.2.3. Sequence Analysis of GAGs

The presence of domains with ordered structure within certain types of GAG chains has prompted the development of strategies for analyzing their precise sugar sequences. The arrangement of sugars relative to a defined reference point, usually the reducing terminal xylose residue, can be assessed either by direct end-labeling or by indirect approaches. In the former case, the reducing end of the chains can be radiolabeled specifically at either the xylitol residue (by treatment of proteoglycan [PG] or peptidoglycan with alkaline $NaB^3H_4$) *(4)* or at the serine residue persisting after exhaustive proteolysis of PG *(16)*.

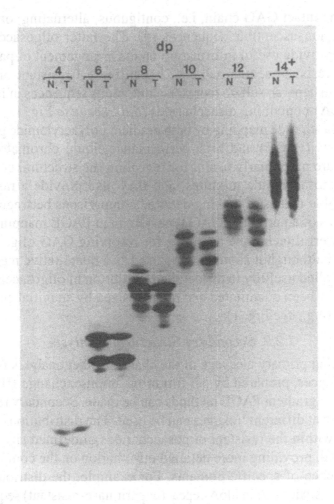

Fig. 1. Oligosaccharide mapping of heparan sulfate oligosaccharides from normal and transformed fibroblasts by gradient PAGE. [3]H-glucosamine-labeled heparan sulfate from normal (N) and transformed (T) mouse fibroblasts in cell culture were treated with heparitinase and the resulting heparitinase-resistant oligosaccharides partially separated by gel filtration on Bio-Gel P6. The individual oligosaccharide peaks (dp4 [tetrasaccharides] to dp ≥ 14 [tetradecasaccharides]) were collected and resolved by gradient PAGE (20–30% gel), transferred to nylon membrane, and detected by fluorography. The oligosaccharide maps show clear differences between the two heparan sulfate species. The predominance of lower mol-wt bands in the oligosaccharides derived from the heparan sulfate from transformed cells suggests a lower content of O-sulfate groups in the heparitinase-resistant IdoA-repeat domains of this species.

Indirect approaches involve "end-referencing" of the chains by coupling the core protein or peptide moieties of PG or peptidoglycan respectively to activated Sepharose *(17–19)*. Specific recovery of oligosaccharides originating at the point of attachment to the protein core (i.e., xylose residue) is then possible after treatment with alkaline borohydride. Alternatively, a hydrophobic moiety can be attached to the reducing terminal xylitol, allowing separation of end-referenced fragments by hydrophobic chromatography *(20)*.

Using the aforementioned strategy, the GAG chains are treated with reagents that cleave at specific linkages. Analysis of the size distribution of the resulting fragments (either specifically end-radiolabeled fragments or those remaining attached to protein, peptide, or hydrophobic moiety for the end-referencing approach) by gel filtration *(17–20)* or gradient PAGE *(16,19)* gives information on the location of the susceptible linkages downstream (i.e., toward the nonreducing end of the chain) from the reducing terminus. Complete digestion defines the position of the first susceptible downstream linkage, whereas partial scission defines the location of linkages further downstream. Studies using the sequencing strategies previously described have shown that the distribution of sugars in heparan sulfate *(4,17,19)*, heparin *(18)*, chondroitin sulfate *(20)*, and dermatan sulfate *(16)* are clearly nonrandom. The further application and development of these strategies should prove to be an important area of future research.

### *1.3. Oligosaccharide Mapping*
### *by Gradient PAGE*

A useful initial method for quantitatively assessing the size and distribution of oligosaccharides derived by cleavage of GAG chains at specific linkages is gel filtration chromatography on BioGel P6 or P10 columns *(5,6,8,9)*. Resolution of fragments up to approx 8–10 disaccharides in length can be obtained, with individual peaks corresponding to oligosaccharides differing in size by one disaccharide unit. Anion-exchange HPLC methods can also be used to map oligosaccharide mixtures *(10–12)*. However, gradient PAGE is a high resolution technique for the separation of GAG oligosaccharides which is superior to gel filtration or anion-exchange HPLC. High resolving power is a result of the combination of a discontinuous buffer system with a pore-size gradient polyacrylamide resolving gel *(7,13)*. Oligosaccharide mapping by gradient PAGE is a rapid and reproducible method for the simultaneous comparison of multiple samples, allowing structural characteristics to be elucidated and compared *(5,7)* (*see* Fig. 1).

## 1.4. Detection

GAG oligosaccharides partially resolved by gel filtration can be detected by monitoring UV absorbance or scintillation counting of radioactivity. Radiolabeled GAG oligosaccharides resolved by gradient PAGE can be transferred and immobilized on nylon membrane using electrotransfer methods, and readily detected by fluorography. Standard commercially available blotting equipment can be used for either "wet" *(5,7)* or "semidry" *(14)* electrotransfer (*see* Note 5 and Chapter 20).

Detection of nonradiolabeled oligosaccharides resolved by gradient PAGE can readily be achieved by direct staining in the gel with cationic dyes if sufficient material is available (approx 100–200 ng/band, 50–100 μg of a complex mixture). Azure A staining is twofold more sensitive than Alcian blue, but increased sensitivity is obtained by a dual-staining method of first staining with Alcian blue, followed by Azure A staining and destaining *(13)*. Oligosaccharides transferred to nylon membrane can also be detected by Alcian blue *(14)*.

A silver-staining method 25- to 50-fold more sensitive than cationic dye staining alone has also been developed *(15)*. This procedure is valuable where the amount of GAG available is very limited and/or where radiolabeling is impractical. Detection of approx 1–2 ng of a single oligosaccharide species can be achieved allowing mapping of a few micrograms of a complex mixture.

## 2. Materials

### 2.1. Sample Preparation

#### 2.1.1. Purification of Glycosaminoglycans

1. DEAE-Sephacel column (1 × 5 cm).
2. Running buffer: 20 m*M* phosphate, pH 6.8.
3. 0.25*M* and 1*M* NaCl in the running buffer.
4. Pronase (1–2 mg/mL) in 100 m*M* Tris-acetate buffer, pH 7.8, containing 10 m*M* calcium acetate.
5. Polysaccharide lyases; e.g., chondroitinase ABC (*see* Table 1).
6. Lyase buffer: *See* supplier's literature for suitable buffer.
7. Sephadex G-25 desalting columns or dialysis tubing.
8. Freeze drier.

### 2.1.2. Depolymerization with Lyases (see Notes 1–4)

1. Specific polysaccharide lyase (*see* Table 1), e.g., for heparan sulfate, use heparitinase I.
2. Lyase buffer solution, e.g., 150 m*M* sodium acetate, 0.1 m*M* calcium acetate, adjusted to pH 7.0 with dilute acetic acid.
3. Unlabeled carrier substrate (for radiolabeled samples), e.g., bovine kidney heparan sulfate.
4. Incubator (suitable for range 25–43°C).

## 2.2. Electrophoresis

1. Vertical slab gel electrophoresis system and power supply (*see* Chapter 19).
2. Electrophoresis buffer: 25 m*M* Tris, 192 m*M* glycine, pH 8.3.
3. Acrylamide and buffer stock solutions (*see* Table 2).
4. Low volume gradient mixing apparatus and peristaltic pump.
5. Microsyringe for sample application.
6. Marker sample: bromophenol blue and phenol red (both about 0.002% [w/v]) in 10% (v/v) glycerol.

## 2.3. Detection

### 2.3.1. Electrotransfer and Fluorography

1. Electrotransfer tank and transfer cassette (*see* Note 5 and Chapter 20).
2. Transfer buffer: 10 m*M* Tris-acetate, pH 7.9, 0.5 m*M* EDTA.
3. Positively charged nylon-66 membrane (e.g., Zetaprobe, Biotrace RP, Biodyne B, Hybond N⁺).
4. Whatman 3MM filter paper.
5. Fluorography materials (*see* Note 6).

### 2.3.2. Alcian Blue / Azure A
### Staining of Nonradiolabeled Oligosaccharides

1. 0.5% (w/v) Alcian blue in 2% (v/v) glacial acetic acid.
2. 2% (v/v) Glacial acetic acid.
3. 0.08% (w/v) Azure A in distilled water.

### 2.3.3. Silver Staining of Submicrogram Quantities of Oligosaccharides (see Note 7)

1. 0.08% (v/v) Azure A.
2. Ammoniacal silver solution: Add gradually 10 mL of fresh 10% (w/v) silver nitrate to a stirred solution of degassed distilled water containing 11.3% (w/v) NaOH and 1.73 mL 35% (w/v) aqueous ammonia. Make this up fresh.

Table 2
Gradient PAGE—Composition of Solutions for Gel Preparation[a]

| | Resolving gel concentration[b] | | |
| --- | --- | --- | --- |
| | Upper limit, T30%/C5% | Lower limit, T20%/C0.5% | Stacking gel T5%/C0.5% |
| Acrylamide (Stock A)[c] | 9.6 mL | — | — |
| Acrylamide (Stock B)[c] | — | 10.65 mL | 3.33 mL |
| 2M Tris-HCl, pH 8.8 | 3.0 mL | 3.0 mL | — |
| 1M Tris-HCl, pH 6.8 | — | — | 2.5 mL |
| Glycerol | 2.5 mL | — | — |
| Double-distilled water | 0.9 mL | 2.3 mL | 13.95 mL |
| Ammonium persulfate (10% [w/v]) | 12.5 µL | 38 µL | 200 µL |
| N,N,N',N'-tetramethylethylene- diamine (TEMED) | 5 µL | 5 µL | 20 µL |
| Total volume | 16 mL | 16 mL | 20 mL |

[a]The recipe given is for a single 20–30% resolving gel approx 27-cm long, 14-cm wide, and 0.75-mm thick (suitable for casting in a 32-cm gel system).

[b]%T refers to the total concentration (w/v) of acrylamide monomer (i.e., acrylamide plus methylene *bis*-acrylamide); %C refers to the concentration of crosslinker (methylene *bis*-acrylamide) relative to total monomer.

[c]Stock A (T50%/C5%): 47.5 g acrylamide and 2.5 g methylene *bis*-acrylamide made up to 100 mL in distilled water. Stock B (T30%/C0.5%): 29.85 g acrylamide and 0.15 g methylene *bis*-acrylamide made up to 100 mL in distilled water.

3. Developing solution: 0.01% (w/v) citric acid, 0.038% (v/v) formaldehyde in degassed distilled water. Make this up fresh.
4. 2.5% (v/v) Acetic acid.
5. Glass staining dishes.

# 3. Methods
## 3.1. Sample Preparation
### 3.1.1. Purification of Glycosaminoglycans

1. Extract proteoglycans or peptidoglycans from tissue or cell culture systems (*see* Note 8 and Chapter 23).
2. Purify solubilized GAGs initially by ion-exchange on DEAE-Sephacel. Load sample in running buffer and wash with 0.25M NaCl (in running buffer) to elute contaminating proteins and hyaluronic acid.
3. Elute the sulfated GAGs with 1M NaCl in the same buffer.
4. Desalt the sample by dialysis or gel filtration on Sephadex G-25; then freeze-dry.

5. Remove unwanted GAG contaminants by differential treatment with specific polysaccharide lyases (*see* Table 1). To purify heparan sulfate, e.g., use chondroitinase ABC treatment to digest chondroitin sulfate, dermatan sulfate, and hyaluronic acid.
6. If not already carried out during extraction, treat the sample with the Pronase solution (*see* Note 8) at 37°C for 24–48 h to produce peptidoglycans.
7. Carry out a final ion-exchange step as in step 2 followed by dialysis and freeze drying to yield peptidoglycans sufficiently pure for further analysis.

### 3.1.2. Depolymerization with Lyases

The following example describes the conditions for depolymerization of heparan sulfate with the lyase heparitinase I. Similar conditions apply to the other heparitinases/heparinases; for the chondroitinases see suppliers literature for optimum buffer and temperature conditions. When degradation methods are used prior to the gradient PAGE mapping, the sample volumes should be kept small (i.e., 20–50 µL) to avoid overloading the gel with salt, particularly in the case of chemical methods. The precise methodological details will depend on the sample.

1. Prepare the sample either freeze-dried or in a small volume of distilled water.
2. Add lyase buffer solution so that the final concentrations are 150 m$M$ sodium acetate and 0.1 m$M$ calcium acetate, pH 7.0.
3. For radiolabeled samples add carrier unlabeled substrate, e.g., bovine kidney heparan sulfate (final concentration, 0.5–1 mg/mL).
4. Add heparitinase I stock enzyme solution (final concentration 10–50 mIU/mL).
5. Incubate at 37°C and monitor the reaction by increase in UV absorbance (*see* Note 1). Under the above conditions near-complete depolymerization is achieved typically in 3–4 h, but digests can routinely be extended to 16 h to ensure cleavage of all susceptible linkages.

## 3.2. Electrophoresis

### 3.2.1. Preparing the Gradient PAGE Gel

1. Assemble the gel unit (according to the manufacturer's instructions and *see* Chapter 19).
2. Prepare and degas the upper and lower limit resolving gel acrylamide solutions without ammonium persulfate and TEMED. (Table 2 gives the details for a standard 20–30% gradient resolving gel. *See* Note 9.)
3. Add ammonium persulfate and TEMED to the upper and lower limit solutions, and place in the mixing and reservoir chambers respectively of the gradient mixing apparatus after thorough mixing.

4. Begin pumping the solution from the mixing chamber into the top of the gel unit between the glass plates (50–100 mL/h), and simultaneously open the valve connecting it to the limit solution chamber. A linear gradient is formed between the glass plates from the bottom (upper limit gel concentration) upward to the top (lower limit gel concentration).

5. When all the gradient has been transferred, overlay the unpolymerized resolving gel with resolving gel buffer. Polymerization should occur (from the top of the gel downward) in approx 1 h. The gel can then be used immediately or stored at 4°C for 1–2 wk.

### 3.2.2. Electrophoresis

1. Immediately before electrophoresis, rinse the resolving gel surface with stacking gel buffer (0.125$M$ Tris-HCl, pH 6.8).

2. Prepare and degas the stacking gel solution, add ammonium persulfate and TEMED, apply to the top of the resolving gel, and immediately insert a well-forming comb between the glass plates.

3. After polymerization (approx 15 min) remove the comb and rinse the wells with electrophoresis buffer.

4. Place the gel unit into the electrophoresis tank and fill the buffer chambers with electrophoresis buffer.

5. Load the oligosaccharide samples (5–50 μL containing approx 10% (v/v) glycerol) carefully into the wells with a microsyringe. Marker samples containing bromophenol blue and phenol red in 10% (v/v) glycerol should also be loaded into separate wells.

6. Run the samples into the stacking gel at 150–200 V for 30 min, followed by electrophoresis at 300–400 V for approx 16 h (for the gel described in Table 2). Heat generated during the run should be dissipated using a heat exchanger with circulating tap water.

7. Terminate electrophoresis when the phenol red marker dye is 2–3 cm from the bottom of the gel (at this point, disaccharides should be 1–2 cm from the bottom of the gel). The bromophenol blue marker dye migrates to a position approximately that of dodecasaccharides (dp 12) derived from heparan sulfate and should be well resolved from the phenol red marker at termination.

## 3.3. Detection

### 3.3.1. Wet Electrotransfer Method (see Notes 5,10, and 11)

1. After termination of electrophoresis remove the stacking gel and equilibrate the resolving gel in transfer buffer for 10 min.

2. Place the gel onto a sheet of filter paper (Whatman 3MM) soaked in transfer buffer, and overlay the gel with a sheet of positively charged

nylon-66 membrane, followed by a further sheet of presoaked filter paper. Ensure good contact between the gel and the nylon membrane by gently rolling a glass tube over the top of the layered assembly.

3. Sandwich the whole assembly between the pads in a transfer cassette and insert into the transfer tank containing transfer buffer precooled to 4°C. The nylon should be on the anodic side of the gel.

4. Transfer at 1.5 V/cm for 3–4 h at 4°C. Most of the marker dyes (particularly the phenol red) should transfer to the nylon. This is a good preliminary visual check on the transfer efficiency.

5. Remove nylon membrane for detection of oligosaccharides by fluorography (*see* Note 6) or by cationic dye staining (*see* Section 3.3.2.).

### 3.3.2. Alcian Blue / Azure A
### Staining of Nonradiolabeled Oligosaccharides (see Note 12)

1. After removal of the gel from the glass plates stain for 15–30 min in 0.5% (w/v) Alcian blue in 2% (v/v) acetic acid.

2. Destain in 2% (v/v) acetic acid.

3. Stain for 15–30 min in 0.08% (w/v) Azure A in distilled water.

4. Destain in distilled water.

5. Oligosaccharides immobilized on nylon membrane can also be detected by staining with Alcian blue (2 min stain, followed by destaining for 2 min).

### 3.3.3. Silver Staining of Submicrogram Quantities
### of Oligosaccharides (see Note 12)

1. Fix the gel in 0.08% Azure A for 15–30 min with gentle agitation.

2. Destain in distilled water. This process is accelerated by adding 5 μL glacial acetic/1 L water.

3. Incubate the gel in freshly prepared ammoniacal silver solution for 1 h.

4. Wash the gel three times (5 min each) with degassed distilled water.

5. Incubate the gel, with constant agitation, in 200 mL of the developing solution. Images usually appear in 1–5 min.

6. Terminate development rapidly when a suitable intensity of staining has been achieved, by incubating in 2.5% (v/v) acetic acid for 30 s.

7. Wash the gel extensively in running tap water until all the cloudiness has been removed. The resulting stained gel is stable and will last indefinitely.

## 4. Notes

1. Polysaccharide lyases are the most commonly used, commercially available enzymes for depolymerization of GAGs. They are bacterial enzymes that cleave specific glycosidic linkages (*see* Table 1) by an eliminative mechanism, resulting in the formation of oligosaccharides with

nonreducing end 4,5-unsaturated uronic acid residues. This UV chromophore provides a convenient method for monitoring the reaction (increase in absorbance at 232 nm). It is thus possible to ensure that depolymerization of all susceptible linkages is complete. When cleaving biosynthetically radiolabeled samples it is advisable to include carrier GAG of an appropriate type in the digest to ensure standard digestion conditions. It is also vital to prepare starting material of sufficient purity in order to avoid difficulties associated with inhibition of enzyme activity.

2. Keratan sulfate can be scissioned with the enzymes keratanase and endo-β-galactosidase, both of which cleave at nonsulfated galactosidic linkages, though keratanase appears to require an adjacent GlcNAc(6S) residue. Hyaluronic acid can be depolymerized by the action of chondroitinases or a number of hyaluronidases from a variety of sources (e.g., *Streptomyces* and leech hyaluronidases).

3. Deaminitive hydrolysis is a useful chemical cleavage method, for the structural analysis of the *N*-sulfated GAGs, heparan sulfate, and heparin. Low pH nitrous acid (pH 1.5) treatment results in specific and near quantitative cleavage at GlcNSO$_3$ residues, the latter being converted into 2,5-anhydromannose, with concomitant chain cleavage and release of SO$_4^{2-}$ and N$_2$ *(21)*. Care needs to be taken in the quantitative interpretation of nitrous acid scission data, particularly in heparin polymers, owing to the occurrence of a ring contraction reaction that precludes chain cleavage *(2,21)*. With unlabeled samples it is possible to radiolabel the products by reduction with NaB$^3$H$_4$ *(1,4)*.

4. Periodate oxidation is an additional chemical method for the depolymerization of heparan sulfate, heparin, and dermatan sulfate. On treatment of heparan sulfate or heparin with periodate at pH 3 and 4°C, GlcA associated with GlcNAc is selectively oxidized, whereas the GlcA, IdoA, and IdoA(2S) residues from *N*-sulfated regions are unaffected. Upon treatment with alkali the oxyheparan sulfate is fragmented into oligosaccharides of general structure GlcNR-(hexuronic acid-GlcNR)$_n$-R where R is the remnant of an oxidized and degraded GlcA residue *(22)*. With dermatan sulfate, nonsulfated IdoA can be selectively oxidized and cleaved under conditions that preclude oxidation of GlcA and IdoA(2S) *(16)*.

5. An alternative electrotransfer method is the "semidry" electrotransfer technique recently reported for GAG oligosaccharides *(14)*. Efficient transfer can be achieved in approx 1–2 h, although small oligosaccharides can be transferred very rapidly (e.g., within 10–15 min).

6. Fluorography of immobilized oligosaccharides can be achieved by thorough drying of the nylon (2 h, 70°C in a vacuum oven), spraying with a

fluorography agent (e.g., En ³Hance from New England Nuclear), air-drying, and exposure to X-ray film (e.g., Kodak X-Omat AR).

7. In silver staining, all electrophoresis and staining reagents must be of the highest purity to minimize background and artifactual staining. Glass dishes and gel plates must be very clean and gels handled with **clean** gloves. Marker dyes must be loaded on to tracks separate from those for samples, as they too can become stained. For ease of handling, small gels are recommended.

8. General procedures for the extraction and purification of proteoglycans from tissues or cell culture systems have been described in detail previously (*24–26* and *see* Chapter 23). Alternatively, GAGs in the form of peptidoglycans can be obtained more directly from fragmented tissue or cell layers by digestion with nonspecific proteases, such as Pronase (*see* Section 3.1.1.). In general, adequately purified peptidoglycan samples are sufficient for structural analysis, although proteoglycans can also be used in some cases. Release of free chains by alkaline borohydride cleavage of the reducing terminal xylose-serine linkage should be avoided if possible, since some release of sulfate groups is known to occur under the conditions generally used.

9. The concentration range of the gradient resolving PAGE gel can be altered to optimize the separation of oligosaccharides of different sizes. A 20–30% gradient is good for general purpose mapping, but improved separation of very small or very large oligosaccharides can be achieved using gel concentrations within the ranges 25–40% and 20–25%, respectively. Different voltage conditions and running times are required for different gradients, and can be established by trial and error.

10. With electrotransfer methods oligosaccharides with low levels of sulfation bind to the membrane with low efficiency *(7)* necessitating care in the quantitative interpretation of data. Although additional layers of nylon can be used to capture fragments that transfer through the first layer, transfer should be assessed and standardized for the particular type of oligosaccharide mixtures under investigation.

11. Oligosaccharides resolved by analytical (or preparative) gradient PAGE and transferred to nylon membrane can be recovered by cutting out specific bands (located by staining or fluorography of parallel strips cut from the membrane edge) and eluting with 2*M* sodium chloride *(14)*. In addition, the semidry transfer technique can be used to purify a major oligosaccharide from minor contaminants; using multiple layers of nylon, oligosaccharide purity increases proportionally with transfer depth *(14)*.

12. A limitation of the cationic dye and silver staining methods which must be borne in mind is the oligosaccharide cut-off imposed by the dye

fixation step. Oligosaccharides with less than a minimum size and/or charge content are not quantitatively fixed within the gel *(15)*. Quantitative analysis of smaller species is best served by gel chromatographic or HPLC techniques.

## Acknowledgments

Thanks to Keiichi Yoshida for provision of heparitinases and to Robert Linhardt, D. Loganathan, and Hui-Ming Wang at the University of Iowa for collaboration on the lyase specificities. The generous support of the Cancer Research Campaign and the Christie Hospital Endowment Fund are acknowledged. Thanks also to Mrs. P. Jones for secretarial assistance.

## References

1. Bienkowski, M. J. and Conrad, H. E. (1985) Structural characterization of the oligosaccharides formed by depolymerisation of heparin with nitrous acid. *J. Biol. Chem.* **260,** 356–365.
2. Guo, Y. and Conrad, H. E. (1989) The disaccharide composition of heparins and heparan sulphates. *Anal. Biochem.* **176,** 96–104.
3. Yoshida, K., Miyauchi, S., Kikuchi, H., Tawada, A., and Tokuyasu, K. (1989) Analysis of unsaturated disaccharides from glycosaminoglycans by HPLC. *Anal. Biochem.* **177,** 327–332.
4. Edge, A. S. B. and Spiro, R. G. (1990) Characterization of novel sequences containing 3-O-sulphated glucosamine in glomerular basement membrane heparan sulphate and location of sulphated disaccharides in a peripheral domain. *J. Biol. Chem.* **265,** 15,874–15,881.
5. Turnbull, J. E. and Gallagher, J. T. (1990) Molecular organization of heparan sulphate from human skin fibroblasts. *Biochem. J.* **265,** 715–724.
6. Maimone, M. M. and Tollefsen, D. M. (1990) Structure of a dermatan sulphate hexasaccharide that binds to heparin cofactor II with high affinity. *J. Biol. Chem.* **265,** 18,263–18,271.
7. Turnbull, J. E. and Gallagher, J. T. (1988) Oligosaccharide mapping of heparan sulphate by polyacrylamide gel electrophoresis and electro-transfer to nylon membrane. *Biochem. J.* **251,** 597–608.
8. Turnbull, J. E. and Gallagher J. T. (1991) Distribution of iduronate-2-sulphate residues in heparan sulphate: evidence for an ordered polymeric structure. *Biochem. J.* **273,** 553–559.
9. Gallagher, J. T. and Walker, A. (1985) Molecular distinctions between heparan sulphate and heparin. *Biochem. J.* **230,** 665–674.
10. Dickenson, J. M., Morris, H. G., Nieduszynski, I. A., and Huckerby, T. N. (1990) Skeletal keratan sulphate chain molecular weight calibration by high-performance gel permeation chromatography. *Anal Biochem.* **190,** 271–275.
11. Rice, K. G., Kim, Y. S., Grant, A. C., Merchant, Z. M., and Linhardt, R. J. (1985) HPLC separation of heparin-derived oligosaccharides. *Anal. Biochem.* **150,** 325–332.

12. Linhardt, R. J., Rice, K. G., Kim, Y. S., Lohse, D. L., Wang, H. M., and Loganathan, D. (1988) Mapping and quantification of the major oligosaccharide components of heparin. *Biochem. J.* **254,** 781–787.
13. Rice, K. G., Rottink, M. R., and Linhardt, R. J. (1987) Fractionation of heparin-derived oligosaccharides by gradient polyacrylamide gel electrophoresis. *Biochem. J.* **244,** 515–522.
14. Al-Hakim, A. and Linhardt, R. J. (1990) Isolation and recovery of acidic oligosaccharides from polyacrylamide gels by semi-dry electrotransfer. *Electrophoresis* **11,** 23–28.
15. Lyon, M. and Gallagher, J. T. (1990) A general method for the detection and mapping of submicrogram quantities of GAG oligosaccharides on polyacrylamide gels by sequential staining with Azure A and ammoniacal silver. *Anal. Biochem.* **185,** 63–70.
16. Fransson, L-A., Havsmark, B., and Silverberg, I. (1990) A method for the sequence analysis of dermatan sulphate. *Biochem. J.* **269,** 381–388.
17. Lyon, M., Steward, W. P., Hampson, I. N., and Gallagher, J. T. (1987) Identification of an extended N-acetylated sequence adjacent to the protein-linkage region of fibroblast heparan sulphate. *Biochem. J.* **242,** 493–498.
18. Rosenfeld, L. and Danishefsky, I. (1988) Location of specific oligosaccharides in heparin in terms of their distance from the protein linkage region in the native proteoglycan. *J. Biol. Chem.* **263,** 262–266.
19. Turnbull, J. E. and Gallagher, J. T. (1991) Sequence analysis of heparan sulphate indicates defined location of N-sulphated glucosamine and iduronate-2-sulphate residues proximal to the protein linkage region. *Biochem. J.* **277,** 297–303.
20. Uchiyama, H., Kikuchi, K., Ogamo, A., and Nagasawa, K. (1987) Determination of the distribution of constituent disaccharide units within the chain near the linkage region of shark cartilage chondroitin sulphate C. *Biochim. Biophys. Acta* **926,** 239–248.
21. Shively, J. E. and Conrad, H. C. (1976) Formation of anhydrosugars in the chemical depolymerisation of heparin. *Biochemistry* **15,** 3932–3942.
22. Fransson, L-A., Sjoberg, I., and Havsmark, B. (1980) Structural studies on heparan sulphate. *Eur. J. Biochem.* **106,** 59–69.
23. Linhardt, R. J., Turnbull, J. E., Wang, H. M., Loganathan, D., and Gallagher, J. T. (1990) Examination of the substrate specificity of heparin and heparan sulphate lyases. *Biochemistry* **29,** 2611–2617.
24. Heinegard, D. and Sommarin, Y. (1987) Isolation and characterization of proteoglycans, in *Methods in Enzymology*, vol. 144 (Cunningham, L. W., ed.), Academic, New York, pp. 319–372.
25. Yanagashita, M., Midura, R. J., and Hascall, V. C. (1987) Proteoglycans: isolation and purification from tissue cultures, in *Methods in Enzymology*, vol. 138 (Ginsberg, V., ed.), Academic, New York, pp. 279–289.
26. Lyon, M. and Gallagher, J. T. (1991) Purification and partial characterization of the major cell-associated heparan sulphate proteoglycan of rat liver. *Biochem. J.* **273,** 415–422.

# CHAPTER 25

# Analysis of Sugar Sequences in Glycoproteins by Glycosidase Digestion and Gel Filtration

*Anthony P. Corfield*

## 1. Introduction

The function of many glycoproteins is directly related to the structure of their oligosaccharide chains *(1–5)*. Subtle changes in oligosaccharide structure involving the linkage of or loss of single monosaccharide units are known to occur in many biological events. Knowledge of this fine structure is necessary if the function of glycoproteins is to be understood.

As many of the oligosaccharide structures are complex, containing different monosaccharides, anomeric linkages, linear and branched forms, and repeating units *(2,3,5)*, it has been important to develop methods that have specificity with regard to the individual monosaccharides and to the anomerity of their linkages. Several different approaches have been applied to this end. These include chemical methods such as gas chromatography and mass spectrometry, nuclear magnetic resonance spectroscopy, and methylation analysis *(2,6,7)*. Complete structural analysis relies on the combination of these approaches for unequivocal results *(2,6,7)*.

All of the methods used to analyze oligosaccharide structure depend on the efficient purification of sufficient material free of contaminants. Such methods have been described for the different types of oligosaccharides found in glycoproteins *(2,8)*. One of the most powerful methods available for sequence analysis is the application of

From: *Methods in Molecular Biology, Vol. 19: Biomembrane Protocols: I. Isolation and Analysis*
Edited by: J. M. Graham and J. A. Higgins  Copyright ©1993 Humana Press Inc., Totowa, NJ

highly purified exoglycosidases of known specificity. In this way information on the nature of the terminal monosaccharide(s) and their linkages can be obtained. The use of glycosidases to solve the structure of unknown oligosaccharides depends on the purity of the glycosidase preparation, a precise knowledge of the specificity of the enzyme and the means to assess the effect of incubation. It is important to test the activity and action of the enzyme preparations used. This can be done using the synthetic *p*-nitrophenyl glycosides, where applicable, in rapid assay of enzyme activity and of contaminating activities *(2,8)*. In addition it is necessary to use oligosaccharides as test substrates in the same way *(2,9,10)*. Thus the efficiency of action can be assessed and the presence of contaminating activities specifically identified.

The use of radiolabeled oligosaccharides has enabled microgram amounts of isolated material to be analyzed by calibrated gel filtration chromatography *(9,10)*. This type of analysis allows the action of glycosidases to be assessed in terms of the type and number of monosaccharides removed in each digestion step *(9,11)*.

The methods described here give an outline for the analysis of oligosaccharides using a range of glycosidases and oligosaccharide standards that are commercially available. Inevitably there are some situations that cannot be accommodated by this method. Thus, for complete analysis, glycosidase digestion might be just one of a number of techniques employed to obtain a final structure. In some cases a glycosidase with the desired specificity may not be available commercially. Although enzyme purification will be an important preliminary step, it is beyond the scope of this chapter to address this problem.

The chapter includes (Section 3.5.4.) an example of how the application of the methods is used to deduce the structure of an oligosaccharide.

# 2. Materials

## 2.1. Radiolabeling of Oligosaccharides

1. 0.1*M* Borate buffer: 0.1*M* boric acid, 0.33*M* NaOH, pH 11.0.
2. 15 m*M* NaB[$^3$H]$_4$ (Amersham International plc., Amersham, UK) in 50 m*M* borate buffer, pH 11.0.
3. 1*M* NaBH$_4$ in 50 m*M* borate buffer, pH 11.0.
4. Glacial acetic acid.
5. Acidified methanol (1–2 drops acetic acid/100 mL).

6. Dowex AG 50WX8, 50–100 mesh (H$^+$ form) (Bio-Rad, Richmond, CA).
7. Whatman 3MM paper (Whatman, Maidstone, UK).
8. Butan-1-ol:ethanol:water, 4:1:1 (v/v/v).

## 2.2. Nonenzymic Removal of Sialic Acid and Fucose

1. 0.02$M$ H$_2$SO$_4$.
2. 0.02$M$ NaOH.

## 2.3. Separation of Acidic and Neutral Oligosaccharides

1. Dowex AG 1X2, 200–400 mesh (Cl$^-$ form) (Bio-Rad).
2. Dowex AG 50WX8, 50–100 mesh (H$^+$ form).
3. 0.5$M$ Pyridine acetate.
4. Chromatography columns (1 × 15 cm and 1 × 5 cm).
5. Freeze drier.

## 2.4. Separation of Oligosaccharides by Gel Filtration

1. BioGel P4 minus 400 mesh (Bio-Rad, Richmond, CA).
2. 2-m Chromatography column (or 2 × 1 m columns joined by 0.5-mm Teflon tubing).
3. High pressure pump for upward elution.
4. Column injector and microsyringe.
5. Dextran: 2 × 10$^5$ mol wt (Pharmacia, Uppsala, Sweden).
6. Glucose oligomers (2–25): Prepare by dissolving 1 g Dextran in 10 mL of 0.1$M$ HCl and heat at 100°C for 4 h under nitrogen. Lyophilize the product and store desiccated at room temperature.
7. Oligosaccharide standards (*see* Table 1 for structures and suppliers).

## 2.5. Glycosidase Digestion (see Section 3.5. for Concentrations and Buffers)

1. *p*-Nitrophenyl derivatives of appropriate glycosides (suppliers listed in Table 2).
2. Glycosidases (*see* Table 2 for specificities and suppliers).
3. Keratan sulfate (bovine cornea) (Boehringer Mannheim, Mannheim, Germany and Sigma Chemical Co., St. Louis, MO).
4. Mannan (baker's yeast) (Sigma Chemical Co.).
5. Buffers for exoglycosidase digestion:
   a. Sodium acetate (50–100 m$M$), pH 4.0–6.0;
   b. Sodium citrate-phosphate (100 m$M$), pH 3.5–6.5;

Table 1
Commercially Available Oligosaccharides Suitable
as Standards and Glycosidase Substrates

| Enzyme/Structure[a] | Supplier[b] |
|---|---|
| Sialidase | |
| Neu5Acα2-3Galβ1-4Glc | A,B |
| Neu5Acα2-6Galβ1-4Glc | A,B |
| Neu5Acα2-6Galβ1-4GlcNAcβ1-2Manα1 | A,B |

Neu5Acα2-6Galβ1-4GlcNAcβ1-2Manα1 ⟍6
Manβ1-4GlcNAcβ1-4 GlcNAc
⟋3
Neu5Acα2-6Galβ1-4GlcNAcβ1-2Manα1

| α-D-Galactosidase | |
|---|---|
| Galα1-4Galβ1-4Glc | B |
| Galα1-3Galβ1-3GlcNAc | B |
| β-D-Galactosidase | |
| Galβ1-3GalNAcβ1-4Galβ1-4Glc | B |
| Galβ1-3GlcNAcβ1-3Galβ1-4Glc | A,B |
| Galβ1-4GlcNAcβ1-2Manα1 | A,B |

Galβ1-4GlcNAcβ1-2Manα1 ⟍6
Manβ1-4GlcNAcβ1-4 GlcNAc
⟋3
Galβ1-4GlcNAcβ1-2Manα1

| N-Acetyl-α-D-Galactosaminidase | |
|---|---|
| GalNAcα1-3GalNAcβ1-3Galα1-4Galβ1-4Glc | B |
| N-Acetyl-β-D-Glucosaminidase | |
| GlcNAcβ1-2Manα1 | A |

GlcNAcβ1-2Manα1 ⟍6
Manβ1-4GlcNAcβ1-4GlcNAc
⟋3
GlcNAcβ1-2Manα1

Fucα1                                                      A
│
GlcNAcβ1-2Manα1 ⟍6          6
GlcNAcβ1-4Manβ1-4GlcNAcβ1-4GlcNAc
⟋3
GlcNAcβ1-2Manα1

| α-D-Mannosidase | |
|---|---|
| Manα1 | A |

Manα1 ⟍6
Manβ1-4GlcNAcβ1-4 GlcNAc
⟋3
Manα1

Table 1 (continued)

| Enzyme/Structure | Supplier |
|---|---|

α-D-Mannosidase (cont'd)

Manα1  
  ＼  
   6  
    Manα1  
      ＼  
       3  
  Manα1  ／  
          6  
           Manβ1-4GlcNAcβ1-4GlcNAc  
          3  
         ／  
Manα1-2Manα1

A,B

β-D-Mannosidase  
  Manβ1-4GlcNAc

B

α-L-Fucosidase  
  Fucα1-2Galβ1-4Glc

A,B

  Fucα1-2Galβ1-3GlcNAcβ1-3Galβ1-4Glc  
                 4  
                 |  
               Fucα1

A,B

  Galβ1-4GlcNAcβ1-3Galβ1-4Glc  
         3  
         |  
       Fucα1

A,B

                        Fucα1  
                          |  
GlcNAcβ1-2Manα1           |  
      ＼                  6  
       6  
  GlcNAcβ1-4Manβ1-4GlcNAcβ1-4GlcNAc  
                 ／ 3  
GlcNAcβ1-2Manα1

A

All glycosidases acting on synthetic substrates  
  *p*-nitrophenyl-glycosides

C–E

[a]Neu5Ac, *N*-acetylneuraminic acid; Gal, galactose; Glc, glucose; GlcNAc, *N*-acetylglucosamine; Man, mannose; GalNAc, *N*-acetylgalactosamine; Fuc, fucose.

[b]Suppliers of oligosaccharides: (A) Oxford Glycosystems, Oxford, UK; (B) BioCarb Chemicals, c/o Accurate Chemicals, New York; (C) Sigma Chemical Co., St. Louis, MO; (D) Genzyme, Boston, MA; (E) Boehringer Mannheim, Mannheim, Germany.

Table 2
Glycosidases Used for Oligosaccharide Analysis

| Enzyme/Source | Buffer[a] | pH Optimum | Supplier[b] | Linkage[c] specificity |
|---|---|---|---|---|
| **Sialidase** | | | | |
| *Vibrio cholerae* | 1 | 5.6 | B,C,F | α2-3>6,8 |
| *Clostridium perfringens* | 1 | 4.5–5.2 | B,C | α2-3>6,8 |
| *Arthrobacter ureafaciens* | 1 | 5.0–5.5 | A–D | α2-6>3,8 |
| Newcastle disease virus | 1 | 5.0–6.0 | A | α2-3>>>6,8 |
| **α-D-Galactosidase** | | | | |
| Green coffee beans | 2 | 6.3–6.5 | A–C | α1-3,6 |
| *Escherichia coli* | 2 | 7.2 | B–E | α1-3,6 |
| **β-D-Galactosidase** | | | | |
| Jack bean | 2 | 3.5 | A–E | β1-6>4>>3 |
| Chicken liver | 3 | 3.6 | A | β1-3,4[d] |
| Bovine testis | 2 | 4.3 | B,C | β1-3>>4 |
| **Endo-β-Galactosidase** | | | | |
| *Bacteroides fragilis* | 1 | 5.8 | A,B,G | -Galβ1-4GlcNAc[d] |
| **N-Acetyl-α-D-Galactosaminidase** | | | | |
| Chicken liver | 2 | 3.6 | A,C,G | α1-3/Ser,Thr |
| **N-Acetyl-β-D-Glucosaminidase** | | | | |
| Chicken liver | 2 | 3.6 | A | β1-2,3,4 |
| Jack bean | 2 | 5.0 | A,C,G | β1-2,3,4,6 |
| *Streptococcus pneumoniae* | 2 | 6.0 | A,B | β1-3,6Gal[d] β1-2(4,6)Man[d] |
| **α-D-Mannosidase** | | | | |
| Jack bean | 1 | 4.5 | A,B,C | α1-2,3,6 Man[d] |
| *Aspergillus phoenicis* | 1 | 5.0 | A | α1-2Man |
| **β-D-Mannosidase** | | | | |
| Snail | 2 | 4.0 | C | β1-4Man |
| **α-L-Fucosidase** | | | | |
| Chicken liver | 2 | 6.0 | A | α1-2,4,6 |
| Bovine testis | 2 | 6.0 | A,C | α1-6>>2,3,4 |

[a]Recommended buffers: (1) 50 m$M$ sodium acetate; (2) 100 m$M$ sodium citrate/phosphate; (3) 50 m$M$ sodium citrate.

[b]Commercial suppliers for enzymes: (A) Oxford Glycosystems, Oxford, UK; (B) Boehringer Mannheim, Mannheim, Germany; (C) Sigma Chemical Co., St. Louis, MO; (D) Seikagaku Kogyo Co., Tokyo, Japan; (E) Miles, Naperville, IL; (F) Behringwerke, Marburg-Lahn, Germany; (G) Genzyme, Boston, MA.

[c]Ser, serine; Thr, threonine (*see* Table 1 for other abbreviations).

[d]*See* Notes 11–18 for more details.

c. Sodium citrate (50 m$M$), pH 4.0–6.0.
Prepare fresh for each series of experiments.
6. 1$M$ Sodium carbonate.

# 3. Methods

## *3.1. Radiolabeling of Oligosaccharides*

1. Dissolve the reducing oligosaccharide (usually <1 mg but may be 10 µg or less) in 0.1 mL borate buffer and add an equal volume of 15 m$M$ NaB[$^3$H]$_4$. Incubate for 90 min, then add 1 vol of 1$M$ NaBH$_4$, and incubate for a further 90 min (*see* Note 1).
2. Destroy excess borohydride by dropwise addition of glacial acetic acid to pH 6.0. Desalt the sample by passage through a column (approx 10 mL) of Dowex 50WX8 (H$^+$ form). Then evaporate repeatedly from acidified methanol (five or six times).
3. Purify the radiolabeled oligosaccharides by paper chromatography on Whatman 3MM paper for 60 h in butan-1-ol:ethanol:water. The oligosaccharides remain at the origin while radioactive contaminants generated during the procedure migrate.
4. Elute the oligosaccharides from the paper with water and freeze-dry.

## *3.2. Nonenzymic Removal of Sialic Acid and Fucose*

Hydrolyze the oligosaccharide in 1 mL of 0.02$M$ H$_2$SO$_4$ at 100°C for 1.5 h. Then neutralize with NaOH. Purify the neutral fraction as in Section 3.3.

## *3.3. Separation of Acidic and Neutral Oligosaccharides (see Notes 2–4)*

1. Dissolve the oligosaccharide fraction in 2 mL distilled water and apply to a column of Dowex AG 1X2 (Cl$^-$ form) Cl$^-$ (1 cm × 10 cm). Wash the column with 20 mL water to recover the neutral fraction.
2. Elute the total acidic fraction with 20 mL of 0.5$M$ pyridine acetate and lyophilize. Dissolve the residue in 1 mL water and pass through a column containing 2 mL of Dowex 50WX8 (H$^+$ form) to remove residual pyridine. Freeze-dry and pass on to analysis by BioGel P4 chromatography (*see* Section 3.4.).

## *3.4. Separation of Oligosaccharides by Gel Filtration*

### *3.4.1. Separation Procedure*

1. Equilibrate the columns at 55°C using a water-jacket or in a warm cabinet.
2. Warm the BioGel P4 minus 400 mesh to 55°C in distilled water and

degas at this temperature. Then load the slurry into the column, maintaining the temperature throughout, and elute the column by upward flow of 0.2–1.0 mL/min with degassed distilled water at 55°C using a high pressure pump.

3. Load samples (0.2–1.0 mL) onto the bottom of the column with an injector and microsyringe, all solutions being prewarmed to 55°C.

4. Monitor the effluent of the column for radioactivity and refractive index using flow-through instruments or by measurement on collected fractions (collect fractions of 0.5–3.0 mL, *see* Note 5)

### 3.4.2. Calibration of the Column

1. Calibrate the column using 0.3–0.4 mg glucose oligomers prepared from the limited acid hydrolysis of Dextran (*see* Section 2.4.). Measure the upper and lower fractionation limits of the column using the unhydrolyzed Dextran and glucose (0.2 mg of each).

2. In addition, calibrate the column with standard oligosaccharides of N- or O-type (*see* Notes 6–8).

3. Express the elution position of each oligosaccharide in glucose units: This allows accurate comparison of reaction products before and after glycosidase digestion.

## 3.5. Glycosidase Digestion

### 3.5.1. Determination of Glycosidase Activity and Purity (see Notes 9 and 10)

1. Make an appropriate enzyme dilution (*see* manufacturer's specifications) in 100 m$M$ sodium citrate/phosphate buffer (except where indicated below) at the optimum pH for the enzyme and add 100 µL to 50 µL of the appropriate *p*-nitrophenyl glycoside (a 10 m$M$ aqueous solution, except where indicated below) and 50 µL of the buffer. Select the incubation conditions from the following:

   a. Sialidase + $N$-acetylneuraminosyl-α2-3-lactose (2 m$M$) at pH 5.1 (*Clostridium perfingens*), pH 4.8 (*Arthrobacter ureafaciens*), pH 5.5 (viral enzymes), or pH 5.5 containing 9 m$M$ CaCl$_2$ and 0.15$M$ NaCl (*Vibrio cholerae*).

   b. α-D-Galactosidase + *p*-nitrophenyl-α-D-galactoside at pH 6.5.

   c. β-D-Galactosidase + *p*-nitrophenyl-β-D-galactoside at pH 3.5.

   d. Endo-β-D-galactosidase + keratan sulfate (3 mg/mL) at pH 5.5. The product is measured as reducing sugar released.

   e. $N$-acetyl-β-D-hexosaminidase + *p*-nitrophenyl-$N$-acetyl-β-D-glucosaminide at pH 4.0–6.0.

   f. $N$-acetyl-α-D-galactosaminidase + *p*-nitrophenyl-$N$-acetyl-α-D-galactosaminide at pH 3.6.

g. α-D-Mannosidase + *p*-nitrophenyl-α-D-mannoside (Jack bean) in 100 mM sodium acetate, 2 mM ZnCl₂ at pH 4.5; an alternative substrate is mannan (2 mM with respect to mannose, for the enzyme from *A. phoenicis*) in 50 mM sodium acetate at pH 5.0.

h. β-D-Mannosidase + *p*-nitrophenyl-β-D-mannoside in 50 mM sodium acetate at pH 4.0.

i. α-D-Fucosidase + *p*-nitrophenyl-α-L-fucoside at pH 6.0.

2. Incubate for 10–15 min at 37°C and terminate the reaction by adding 300 μL 1M sodium carbonate.

3. Measure the absorbance at 400 nm and compare the activity with the quoted values given in the manufacturer's information sheet.

4. Test for the presence of low levels of contaminating glycosidase activities by incubating the enzyme with other *p*-nitrophenyl glycosides using the same buffer conditions but with 5- to 10-fold more enzyme and for times of 60 min and longer.

### 3.5.2. Glycosidase Activity
### with Oligosaccharide Standards

1. Incubate a specific radiolabeled oligosaccharide standard (at least 10⁴ counts/min), with enzyme and buffer (*see* Section 3.5.3.) under optimum conditions for 18–24 h.

2. Mix the products with glucose oligomers and chromatograph on BioGel P4 (*see* Section 3.4.).

3. Compare the elution position with that of the expected product and note the extent of reaction and the presence of other products on BioGel P4; these indicate activity and/or specificity.

4. Repeat the same procedure with other standard oligosaccharides to check for contaminating glycosidase activities. It is important to check for the presence of contaminating enzymes using the oligosaccharide substrates as well as with the *p*-nitrophenol glycosides.

### 3.5.3. Analysis of Oligosaccharides
### by Glycosidase Digestion

1. Digest the radiolabeled oligosaccharide (10–50 × 10⁴ counts/min) sequentially with glycosidases at 37°C for 18–24 h (for enzyme conditions *see below*). Then use BioGel P4 chromatography (*see* Section 3.4.) to evaluate the result of each incubation.

2. Use the product of each incubation as the substrate for the next digest. Incubate each glycosidase at its own optimal conditions to yield maximum information on the structure of the oligosaccharide under study.

a. Sialidases: Substrate (10–20 μM) in 50 mM sodium acetate pH 5.1 (*Clostridium perfingens*), pH 4.8 (*Arthrobacter ureafaciens*), pH 5.5

(viral enzymes), or pH 5.5 containing 9 m$M$ CaCl$_2$ and 0.15$M$ NaCl (*Vibrio cholerae*). Incubate bacterial enzymes at 1–2 U/mL and viral enzymes at 0.5 U/mL (*see* Notes 11 and 12).

  b. α-D-Galactosidase: Substrate (30 μ$M$) in 100 m$M$ sodium citrate/phosphate pH 6.0 with 5 U/mL enzyme.

  c. β-D-Galactosidases: Substrate (30 μ$M$) in 100 m$M$ sodium citrate/phosphate pH 4.0 with 5 U/mL enzyme. Use the enzymes from Jack bean and bovine testis to discriminate between β1-3 and β1-4 linkages. Incubate the Jack bean enzyme at <1 U/mL with a substrate concentration of 15 m$M$ to cleave β1-4 links only. The bovine testis enzyme at <1 U/mL and 20–30 μ$M$ substrate concentration cleaves β1-3 linkages only (*see* Note 13).

  d. Endo-β-galactosidase: Substrate (20 μ$M$) in 50 m$M$ sodium acetate pH 5.8 containing 250 μg/mL bovine serum albumin with 0.2 U/mL enzyme (*see* Note 14).

  e. *N*-acetyl-α-D-galactosaminidase: Substrate (20 μ$M$) in 100 m$M$ sodium citrate/phosphate buffer pH 4.0 with 1–2 U/mL enzyme.

  f. *N*-acetyl-β-D-glucosaminidases: Substrate (20 μM) in 100 m$M$ sodium citrate/phosphate pH 4.0 (chicken liver) or pH 5.0 (Jack bean) at 10 U/mL enzyme. Use <10 U/mL (*Streptococcus pneumoniae*) enzyme in 100 m$M$ sodium citrate/phosphate at pH 6.0 to obtain specific cleavage of β1-2Man linkages in *N*-linked oligosaccharides (*see* Note 15).

  g. α-D-Mannosidases: Substrate (20 μ$M$) in 100 m$M$ sodium acetate. 2 m$M$ ZnCl$_2$, pH 5.0 with 10–100 U/mL enzyme (Jack bean) or in 100 m$M$ sodium citrate/phosphate pH 5.0 with 2 U/mL enzyme (*Aspergillus phoenicis*). Make a second addition of the Jack bean enzyme after 12 h. Use the Jack bean enzyme at 6 U/mL to discriminate between α1-3 and α1-6 linked terminal mannose residues (*see* Note 16).

  h. β-D-Mannosidase: Substrate (30 μ$M$) in 50 m$M$-sodium acetate pH 4.0 with 0.3 U/mL enzyme (Snail, *Aspergillus niger, see* Note 17).

  i. α-L-Fucosidases: Substrate (30 μ$M$) in 100 m$M$ sodium citrate/phosphate pH 6.0 buffer with 2 U/mL enzyme (*see* Note 18).

### 3.5.4. Analytical Strategy for the Analysis of the Oligosaccharide Structure of Horse Pancreas Ribonuclease

The determination of oligosaccharide structure commences with the isolation of the oligosaccharide(s) from the glycoprotein *(2,4,8,11)* and radiolabeling as described in Section 3.1. This product is then taken sequentially through the procedures in Sections 3.2.–3.5.

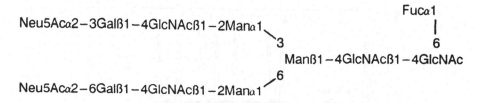

Fig. 1. Structure of oligosaccharide. Neu5Ac, *N*-acetylneuraminic acid (sialic acid); Gal, galactose; GlcNAc, *N*-acetylglucosamine; Man, mannose; Fuc, fucose.

To illustrate the methodological approach, an example is given for the oligosaccharide from horse pancreatic ribonuclease *(18)*. The structure of this oligosaccharide has been reported to be as shown in Fig. 1. This is a typical N-linked, biantennary oligosaccharide containing one fucose residue attached to the chitobiosyl core and two sialic acids with different linkages. The size and anomeric linkage can be determined using glycosidases. The procedure assumes no knowledge of the structure whatsoever. However, the procedure can be simplified if a knowledge of N-linked oligosaccharide structure is assumed and the expected terminal sugars are digested with the appropriate glycosidase at each step.

1. Section 3.3. (ion-exchange) shows that the oligosaccharide has a negative charge and there is no neutral material. Subsidiary chemical analysis shows the absence of sulfate or phosphate, so the charge is caused by sialic acid only.
2. Section 3.4. (gel filtration on BioGel P4) gives a single peak near $V_0$ (Fig. 2).
3. Analysis of the terminal monosaccharides using the methods in Sections 3.5.3. and 3.4.; *see* Fig. 2a and Table 3. Results (Table 3) show that terminal sialic acid and fucose are present.
4. Digestion with bacterial sialidase (*see* Sections 3.5.3. and 3.4.) completely removes sialic acid and changes the BioGel P4 elution profile (*see* Fig. 2b,c).
5. After digestion with viral sialidase (*see* Sections 3.5.3. and 3.4.) the product is still acidic and still elutes near $V_0$ on BioGel P4. Conclude that an α2-6 linked sialic acid is present (*see* Fig. 2b,c).
6. Sequential digestion with exoglycosidases followed by gel filtration (*see* Sections 3.5.3. and 3.4.) is carried out according to Table 4. The elution profiles after each digestion (Fig. 2a) indicate a biantennary structure with fucose (*see* Fig. 3).

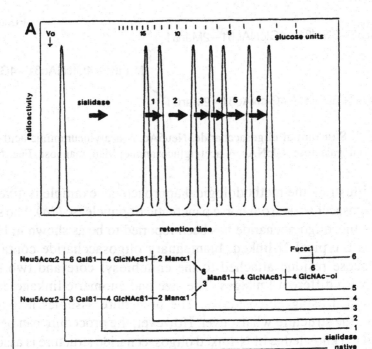

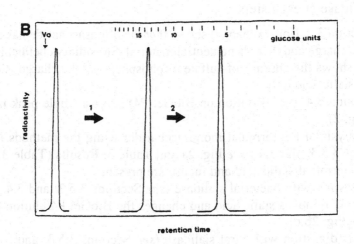

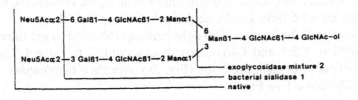

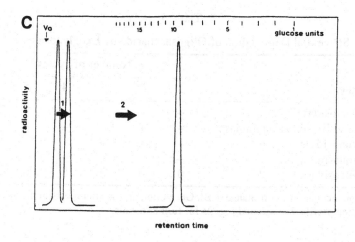

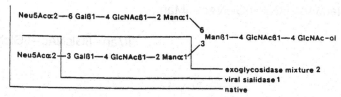

Fig. 2. *(opposite page and above)* Oligosaccharides structure by glycosidase digestion and gel filtration. (A) Stepwise degradation of the completely desialylated oligosaccharide as indicated in Table 3. The results of each digest are shown together with a fragmentation scheme based on the results of each experiment. (B) Digestion with bacterial sialidase and exoglycosidase mixture. (C) Digestion with viral sialidase and exoglycosidase mixture. The bacterial sialidase (B) cleaves both α2-3 and α2-6 linked sialic acids; the viral enzyme (C) only α2-3 linkages. That both types of linkage are present in the same molecule is shown using a mixture of glycosidases acting on the desialylated structures as far as the βMan core residue. The BioGel P4 profiles and fragmentation schemes are shown for each experiment.

Table 3
Determination of Terminal Monosaccharides

| Enzyme treatment | Change in BioGel P4 profile |
| --- | --- |
| Sialidase | + |
| α-D-Galactosidase | − |
| *N*-Acetyl-α-D-galactosaminidase | − |
| α-Fucosidase (after sialidase) | + |

Table 4

Sequential Degradation of Oligosaccharide by Exoglycosidases

| Enzyme | Number of cycles | | | | | |
|---|---|---|---|---|---|---|
|  | 1 | 2 | 3 | 4 | 5 | 6 |
| β-Galactosidase | + | – | – | – | – | 0 |
| *N*-Acetyl-β-D-glucosaminidase | – | + | – | – | + | 0 |
| α-Mannosidase | – | – | + | – | – | 0 |
| β-Mannosidase | – | – | – | + | – | 0 |
| α-L-Fucosidase | 0 | 0 | 0 | 0 | 0 | + |

0, no incubation; –, no change in BioGel elution profile after incubation; +, change in elution profile. *See text* and Fig. 2a for further details.

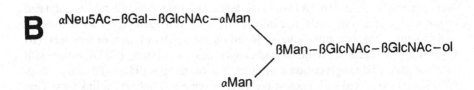

Fig. 3. Discrimination of mannose linkages. For details, *see* Section 3.5.4.

7. The determination of the presence of both α2-3 and α2-6 linked sialic acid achieved using the products of the specific action of viral and bacterial sialidase (*see* Section 5.2., steps 4 and 5), after α-L-fucosidase digestion.

8. The products are treated with a mixture of exoglycosidases (β-galactosidase, *N*-acetyl-β-D-glucosaminidase, and α-mannosidase), which degrade only nonsialylated oligosaccharide branches to the core trisaccharide. Any sialic acid remaining on one branch protects the Galβ1-4GlcNAcβ1-2Man arm against degradation giving a product with a characteristic elution profile on BioGel P4.

9. Complete desialylation of the products of exoglycosidase action with bacterial sialidase allows easier identification on BioGel P4 (*see* Note

5). The results for both sialidase digestions are shown in Fig. 2b,c. Only the bacterial sialidase allows the exoglycosidase mixture to produce the core trisaccharide.

10. The action of Jack bean α-mannosidase can be used to discriminate between α1-3 and α1-6 linked mannose in partially degraded oligosaccharide structures (*see* Note 14 and ref. *16*). Digestion at 6 U/mL will release α1-3 linked mannose but not α1-6. Thus digestion of the viral sialidase product gives the structure in Fig. 3a and not Fig. 3b, showing that the α2-6 linked sialic acid is probably located on the α1-6 mannose arm.

## 4. Notes

1. Radiolabeling may be more conveniently carried out during the purification procedure. This has advantages where a mixture of oligosaccharides are present in one glycoprotein and may be cleaved and labeled at the same time e.g., alkaline β-elimination and reduction.

2. The separation of acid and neutral oligosaccharides can be carried out on a wide range of HPLC chromatography systems including reverse-phase and ion-exchange (*8*) in addition to the gel filtration method described here. Whichever system is used, calibration is required to allow the effect of glycosidase digestion to be assessed in terms of individual or multiple monosaccharide loss.

3. In some cases it may be necessary to remove all sialic acid and fucose residues and analyze the neutral oligosaccharides. This may arise when terminal structures are present which cannot be degraded by available glycosidases as with, e.g., fucose residues in some Lewis blood group antigens and with sialic acid linked in ganglioside $G_{MI}$. Alternatively it can be used to determine whether a mixture of oligosaccharides differ only in their peripheral fucose and sialic acid residues but possess the same basic neutral core structure. In these cases the total complement of sialic acid and fucose is hydrolyzed with mild acid.

4. Other acidic groups may be present in oligosaccharides, sulfate being more common than phosphate. In these cases it is important to eliminate sialic acid enzymically and to recheck for negative charge on ion-exchange chromatography. Determination of sulfate and phosphate by chemical means (*8*) may be necessary for absolute structural determination as both groups may block some glycosidase steps.

5. BioGel P4 shows the best fractionation range for oligosaccharides. However in the case of some larger oligosaccharides, and in particular those containing sialic acid, it may be necessary to use BioGel P6. Gel filtration is carried out as described for BioGel P4 and the column is calibrated with glucose oligomers.

6. Standard oligosaccharides of N- and O-type are available commercially (*see* Table 1). These can be radioactively labeled as described and are used to confirm elution positions in glucose units on BioGel P4 and also as specific substrates to test the activity of glycosidase action.

7. Mix radioactive samples with glucose oligomers to improve the accuracy of analysis. Run samples (without glucose oligomers) that will be further analyzed by other methods relying on carbohydrate analysis.

8. The BioGel P4 column can be used repeatedly and gives very reproducible results. The purity of the samples applied governs the working life of the column. Regular calibration with glucose oligomers and a standard radioactive oligosaccharide is used to monitor the reproducibility and viability of the column.

9. There are a range of exoglycosidases available commercially (see Table 2). The purity of these preparations varies depending on the enzyme source and the supplier. It is important to test the preparations with respect to contamination by other glycosidase activities. This is done using both *p*-nitrophenyl glycosides and natural substrates (oligosaccharides). The specificity of the different enzyme preparations must be carefully noted where it is given, as some glycosidases act on oligosaccharides but not on synthetic substrates and others act conversely *(8)*.

10. The units of enzyme activity are measured with defined substrates under standard conditions such that 1 U of activity is the amount of enzyme that will release 1 μmole of monosaccharide product from substrate/minute at optimum pH and 37°C.

11. Sialidases from influenza or Newcastle disease viruses are used to hydrolyze α2-3 linked sialic acid at the enzyme concentration given. Under these conditions only limited hydrolysis of α2-6 linkages occurs *(12)*.

12. Sialic acid substitution influences the rate of reaction of sialidases. The presence of *O*-acetyl groups reduces the rate depending on the site and degree of substitution *(13)*. Not all commercially available sialidases will release 4-*O*-acetylated sialic acids, whereas 7-9 *O*-acetylation causes a reduction in hydrolysis, which increases from mono- to tri-*O*-acetylated derivatives. The *O*-acetyl esters can be readily removed by saponification in 50 m*M* NaOH at 4°C for 60 min. The influence of *N*-glycolylneuraminic acid in reducing the rate of sialidase action is small and can be overcome at the normal enzyme concentrations at the extended incubation times (18–24 h) described.

13. The β–galactosidases show specificity for the nature of the partner sugar (e.g., GlcNAc or GalNAc) in anomeric linkage with galactose in addition to the type of linkage (β1-3,4 or 6). The specificity of the enzyme used should be checked as described. Note that the Jack bean enzyme

cleaves Galβ1-6GlcNAc more efficiently than Galβ1-4GlcNAc, with Galβ1-3GlcNAc being the poorest substrate of the three *(14)*.

14. Endo-β-galactosidases cleave within the oligosaccharide chain and are specific for the sequence -GlcNAcβ1-3Galβ1-4GlcNAc/Glc; where the β1-4 linkage is hydrolyzed. This sequence is commonly found in the poly-*N*-acetyllactosamine repeat units of some glycoproteins *(15)*. Incubation under the conditions described leads to a significant reduction in the size of the oligosaccharides containing such repeat units and are readily seen as such on BioGel P4 chromatography.

15. The *N*-acetyl-β-D-hexosaminidase from *S. pneumoniae* shows specificity for β1-2Man linkages at low enzyme concentrations provided the mannose is not substituted at position 6. Further, the presence of a bisecting β1-4 linked GlcNAc results in hydrolysis of β1-2 linked GlcNAc only on the α1-3 mannose branch. The same enzyme at higher concentration, and other enzymes cleave β1-2,3,4 and 6 linkages. Jack bean and *S. pneumoniae* enzymes also act on GlcNAc β1-3 and 6 in *O*-linked oligosaccharides *(8)*.

16. The α-mannosidase from *Aspergillus phoenicis* has a strict specificity for α1-2 linkages and is thus a valuable tool for the detection of mannose linked in this manner in high mannose oligosaccharides.

17. Although a commercial source of β-mannosidase is available, significant amounts of *N*-acetyl-β-D-hexosaminidase and α-D-mannosidase are often present, hence the validity of results may be affected. Other sources, notably from *Aspergillus niger (16)* have been purified and used.

18. The analysis of fucose residues in *O*-linked oligosaccharides by α-fucosidase action is limited because of the incomplete information available for these substrates and the absence of commercial sources of suitable enzyme. In particular the cleavage of α1-3GlcNAc linkages is difficult. The substrate specificities of α-fucosidases are complex and need careful attention with defined substrates.

## References

1. Rademacher, T. W., Parekh, R. B., and Dwek, R. A. (1988) Glycobiology. *Ann. Rev. Biochem.* **57,** 785–838.
2. Montreuil, J. (1982) Glycoproteins, in *Comprehensive Biochemistry,* vol. 19B, part II (Neuberger, A. and van Deenen, L. L. M., eds.), Elsevier, Amsterdam, pp. 1–188.
3. Kornfeld, R. and Kornfeld, S. (1985) Assembly of asparagine-linked oligosaccharides. *Ann. Rev. Biochem.* **54,** 631–664.
4. Kobata, A., Amano, J., Mizochi, T., and Endo, T. (1990) Altered glycosylation of human chorionic gonadotropin in trophoblastic diseases and its use for the diagnosis of choriocarcinoma, in *Glycobiology*, vol. 3 (Welply J. K. and Jaworski, E., eds.), Wiley-Liss, New York, pp. 115–126.

5. Carlstedt, I., Sheehan, J. K., Corfield, A. P., and Gallagher, J. T. (1985) Mucous glycoproteins: a gel of a problem. *Essays Biochem.* **20,** 40–76.
6. Vliegenthart, J. F. G., Dorland, L., and van Halbeek, H. (1983) High resolution ¹H-nuclear magnetic resonance spectroscopy as a tool in the structural analysis of carbohydrates related to glycoproteins. *Adv. Carbohydr. Chem. Biochem.* **41,** 209–374.
7. Dell, A. (1987) F.a.b.- mass spectrometry of carbohydrates. *Adv. Carbohydr. Chem. Biochem.* **45,** 20–72.
8. Montreuil, J., Bouquelet, S., Debray, H., Fournet, B., Spik, G., and Strecker, G. (1986) Glycoproteins, in *Carbohydrate Analysis. A Practical Approach* (Chaplin, M. F. and Kennedy, J. F., eds.), IRL at Oxford University Press, Oxford, UK, pp. 143–204.
9. Yamashita, K., Mizuochi, T., and Kobata, A. (1982) Analysis of oligosaccharides by gel filtration. *Methods Enzymol.* **83,** 105–126.
10. Kobata, A., Yamashita, K., and Takasaki, S. (1987) BioGel-P4 column chromatography of oligosaccharides: effective size of oligosaccharides expressed in glucose units. *Methods Enzymol.* **138,** 84–94.
11. Parekh, R. B., Dwek, R. A., Sutton, B. J., Fernandes, D. L., Leung, A., Stanworth, D., Rademacher, T. W., Mizuochi, T., Taniguchi, F., Nagano, Y., Miyamaoto, T., and Kobata, A. (1985) Association of rheumatoid arthritis and primary osteoarthritis with changes in the glycosylation pattern of total serum IgG. *Nature* **316,** 452–457.
12. Corfield, A. P., Wember, M., Schauer, R., and Rott, R. (1982) The specificity of viral sialidases. The use of oligosaccharide substrates to probe enzymic characteristics and strain specific differences. *Eur. J. Biochem.* **124,** 521–525.
13. Corfield, A. P., Sander-Wewer, M., Veh, R. W., Wember, M., and Schauer, R. (1986) The action of sialidases on substrates containing *O*-acetylated sialic acids. *Biol. Chem. Hoppe-Seyler* **367,** 433–439.
14. Li, S-C., Mozzotta, M. Y., Chien, S-F., and Li, Y. T. (1975) Isolation and characterization of Jack Bean β-galactosidase. *J. Biol. Chem.* **250,** 6786–6791.
15. Scudder, P., Hanfland, P., Uemura, K-I., and Feizi, T. (1984) Endo-β-D-galactosidase of *Bacteroides fragilis* and *Escherichia freundii* hydrolyse linear but not branched oligosaccharide domains of glycolipids of the neolacto series. *J. Biol. Chem.* **259,** 6586–6592.
16. Bouquelet, S., Spik, G., and Montreuil, J. (1978) Properties of a β-mannosidase from *Aspergillus niger. Biochim. Biophys. Acta* **522,** 521–530.

# CHAPTER 26

# The Extraction and Analysis of Glycosphingolipids

## Norman A. Gregson

## 1. Introduction

### 1.1. Background

Glycolipids are an important group of diverse molecules present in the plasma membrane and to a more limited extent in the intracellular membranes of the Golgi and lysosomal systems of eukaryotic cells. They are prominent as antigenic determinants and as the ligands for specific lectin binding, and there are many examples in which they have been used as cell specific markers. The cellular glycolipid pattern frequently shows changes with the state of differentiation and may be altered following viral transformation. In animal cells the most common and most varied glycolipids are glycosphingolipids, and the methods of isolation and analysis given are mainly for this class of molecule. Methods are given for the extraction of total glycolipid from small volumes of tissue and for the isolation of the glycolipids from such extracts. Methods of fractionating and quantitating the acidic and neutral glycolipids are described, which do not require the use of specialized equipment such as high-performance liquid chromatography (HPLC), gas chromatography (GC), or mass spectroscopy and which can be performed easily in the general laboratory.

### 1.2. Estimation of Major Lipid Classes in Fractions

As the neutral and acidic glycolipids from animal tissues are predominantly glycosphingolipids, they can be quantified by measurement of the sphingosine content. The colorimetric procedure of Lauter

From: *Methods in Molecular Biology, Vol. 19: Biomembrane Protocols: I. Isolation and Analysis*
Edited by: J. M. Graham and J. A. Higgins Copyright ©1993 Humana Press Inc., Totowa, NJ

and Trams *(1)* relies on the estimation of free sphingosine base following acid hydrolysis. The procedure is suitable for the measurement of 0.01–0.1 μmol of sphingosine, but can be scaled down for the estimation of smaller quantities.

The most common acidic glycolipids are those containing *N*-acetyl neuraminic acid (NANA), and most colorimetric methods rely on reaction with resorcinol for their quantitation. The modification introduced by Jourdian et al. *(2)* involves a differential oxidation step using periodic acid. It increases the sensitivity of the procedure and removes interference by other sugars and the acetate ion. It can be conveniently scaled down to increase further the sensitivity. It slightly underestimates the NANA content of the higher gangliosides so a standard of whole brain ganglioside mixture is best used, in addition to a standard of free NANA. The additional acidic glycolipids are the sulfate or sulfonated esters and these can be quantified using a dye-partitioning procedure *(3)*.

### 1.3. Analysis of Glycolipids
### by Thin-Layer Chromatography

Thin-layer chromatography (TLC) on silica gel layers provides the simplest and the first step in the qualitative analysis of tissue glycolipids. The use of high-performance TLC (HPTLC) plates further increases the resolving power and sensitivity of detection. The basic procedures of plate activation, sample application, and performance of the separation are as used in general TLC procedures (*see* Chapter 14).

Detection and identification of the individual glycolipids is routinely achieved by chemical staining and by comparison with known standards. For the more specific recognition and the characterization of glycolipids on the TLC plate, their reactivity with specific antibodies and lectins can be used either directly or after treatment with glycosidases. In general, such detection methods are more sensitive than chemical staining procedures by one to two orders of magnitude. The procedure given in Sections 2.3.2. and 3.3.2. is for the use of antibody, but essentially the same procedure is used for lectin binding and for the treatment with glycosidases. Further information about the glycolipids can be obtained on individual purified lipids by methods that are general to carbohydrate analysis (Chapters 24 and 25).

## 2. Materials

The materials and methods are given for about 1 g of tissue, but they are generally applicable to any amount of animal tissue or cultured cells. If only small amounts of tissue are used, only the major glycolipids may be detected; obviously with cultured cells the use of radiolabeled metabolic precursors can be used to enhance detection. For a more complete analysis including the minor constituents, much larger amounts of tissue must be used, usually in the range of 2–10 g depending on the tissue type.

### 2.1. Tissue Extraction and Preparation of "Neutral" Glycolipid and Ganglioside Fractions

1. Chloroform and methanol (analytical grade) for the following solvent mixtures: chloroform:methanol (C:M), 2:1 (v/v); C:M, 1:1 (v/v); and theoretical upper phase (TUP) C:M:0.1$M$ KCl, 3:48:47 (v/v/v).
2. Pyridine, acetic anhydride, toluene, ethyl acetate, hexane, dichloroethane, and acetone, all dry and of analytical grade.
3. Prepare 0.5% sodium methoxide by the careful addition of 0.3 g of fresh sodium metal to 100 mL dry methanol. **Caution:** This must be done with the methanol behind an explosion screen in a fume cupboard.
4. Florisil, activated magnesium silicate for chromatography, 100–200 mesh. Before use, prewash the gel with 8 vol of water on a Buchner funnel; dry for 6–8 h at 100°C, and store in a stoppered bottle.
5. Glass solvent resistant chromatography column, approx 0.7 × 30 cm fitted with a solvent reservoir.
6. Small (3 mL) C18 reversed-phase sample preparation columns, such as those supplied by Anachem (Luton, Beds., UK).
7. Convert DEAE-Sephadex A25 into the acetate form by suspending in 10 vol of C:M:0.8$M$ ammonium acetate. 30:60:8 (v/v/v). Let the resin stand for 30 min before filtering. Repeat the treatment three times, then suspend in the same solvent mixture once more and leave overnight at 4°C. After filtering, wash the gel three times by suspension in 10 vol of C:M:water, 30:60:8 (v/v/v).
8. Homogenizer, either glass with glass pestle or a stainless steel high-speed blender of the Polytron type suitable for dealing with 10–20 mL.
9. Glass scinter filter, 20-mm diameter, with appropriate glass fiber filters, fitted to either a small filter flask or filter tube (150 × 24 mm) for attaching to a vacuum line.
10. Glass centrifuge tubes, 30–50 mL, and a low-speed centrifuge with appropriate swing-out rotor.

11. Rotary evaporator.
12. Desiccator with phosphorus pentoxide.
13. 0.1$M$ KCl.

## 2.2. Estimation of Major Lipid Classes in Fractions

### 2.2.1. Estimation of Total Glycosphingolipid

1. Ethyl acetate and dioxane, both analytical grade.
2. 1$M$ $H_2SO_4$, 24% (w/v) NaOH, 0.1$M$ sodium acetate buffer, pH 3.6.
3. Methyl orange, 100 mg/100 mL water. Wash the solution five times with 5 mL of chloroform before use.
4. A standard solution of galactocerebroside (Sigma), 1 mg/mL in C:M, 1:1, diluted 1/10 just before use (0.122 µmol/mL).
5. Low-speed centrifuge with appropriate swing-out rotor.
6. Glass-stoppered tubes (10 mL).

### 2.2.2. Estimation of N-Acetylneuraminic Acid

1. 5-mL Stoppered Pyrex tubes, or 1.5-mL Eppendorf tubes with the caps pierced with a fine needle.
2. 0.04$M$ Periodic acid: Prepare a stock solution of 0.4$M$ and store in aliquots at –20°C.
3. Resorcinol: Recrystallize from benzene and dry. A stock solution, 6 g/100 mL in water, can be stored at 4°C for 4–6 wk.
4. 0.1$M$ $CuSO_4$.
5. HCl: Prepare a 28% (w/w) solution, using analytical grade conc. HCl.
6. Working resorcinol reagent: Prepare fresh by making 10 mL of stock aqueous 6% (w/v) resorcinol, 30 mL of water, 60 mL of 28% HCl, and 0.25 mL 0.1$M$ $CuSO_4$.
7. Tertiary butyl alcohol (2-methyl-propan-2-ol).
8. N-acetylneuraminic acid (NANA): Prepare a stock standard solution of 0.2 mg/mL (approx 0.65 µmol/mL); store in aliquots at –20°C.
9. Standard mixture of bovine whole brain gangliosides (Sigma type IV, approx 30% NANA, average composition of disialoganglioside): Prepare stock solution of 0.5–0.6 mg/mL in methanol, store at –20°C.
10. Visible wavelength spectrophotometer.

### 2.2.3. Estimation of Sulfated Glycolipids

1. 25 m$M$ $H_2SO_4$.
2. Azure A solution: Dissolve 40 mg in 5 mL 25 m$M$ $H_2SO_4$, make up to 100 mL with water, filter through glass-fiber paper. The dye solution can be stored in a dark glass bottle for up to 1 wk at room temperature.
3. A standard solution of sulfatide (galactocerebroside-3-sulfate), 0.05 µmol/mL in C:M, 1:1.

4. 12-mL Glass-stoppered tubes.
5. Visible wavelength spectrophotometer.

## 2.3. Analysis of Glycolipids by Thin-Layer Chromatography

### 2.3.1. By HPTLC and Chemical Detection

1. Standard HPTLC plates, silica gel G60, 10 × 10 cm. Plates with a preconcentrating zone are useful.
2. Small chromatography tanks, those supplied by Camag (BDH-Merck Ltd.) have the bottom divided into two sections, which is convenient for the preconditioning of the plate.
3. Solvents for developing the chromatograms: Nearly all of those commonly used for glycolipids are based on chloroform, methanol, water (CMW) mixtures. All solvents should be of analytical grade. There are a large number of solvent mixtures in the literature suitable for the resolution of glycolipid mixtures, but the most common mixtures used for the separation of gangliosides are C:M:0.5% $CaCl_2$, 60:40:9 (v/v/v) and C:M:2.5$M$ $NH_4OH$, 60:35:8 (v/v/v), and for neutral glycolipids C:M:W, 60:35:8 (v/v/v) is the most widely used.
4. Glass syringes, 10–50 μL, fitted with a flat-ended needle are best for sample application.
5. Hot plate, or oven for heating plates after spraying.
6. Spray/atomizer for application of detection reagents. An all glass spray is the best since the most common detection reagents contain either sulfuric acid or HCl. Most of the spray units using plastic spray heads are not stable with the reagents used.
7. Detection reagents: Details are given in Section 3.3.1.
8. Reference standard compounds: Purified individual glycolipids can be bought from specialist companies (e.g., BioCarb), and several companies sell a few selected compounds or standard tissue extracts suitable as standards such as bovine brain gangliosides.

### 2.3.2. Detection by Ligand Binding

1. Aluminum-backed HPTLC, or plastic-backed TLC silica gel plates. These plates have more binder in the layers and so are intrinsically more stable to aqueous buffers. They can be cut with a sharp scalpel and straight edge to any size desired from 20 × 20 cm sheets.
2. Poly-isobutylmethacrylate: A saturated solution in n-hexane.
3. Phosphate buffered saline (PBS): In 1 L dissolve 8.0 g NaCl, 0.20 g KCl, 1.15 g $Na_2HPO_4 \cdot 2H_2O$ and 0.2 g $KH_2PO_4$.
4. 1% Bovine serum albumin in PBS (BSA-PBS).

5. Plastic boxes or trays for exposing TLC strips to antibodies and for washing the same. The strips need not be immersed in antibody solution, it can be pipeted over the surface, about 60 $\mu L/cm^2$. Small plastic stoppers can be stuck onto the base of a plastic box to provide small supporting plinths to prevent the coating solution from being drawn off.
6. Rocker platform (optional).

# 3. Methods

## 3.1. Tissue Extraction and Preparation of Neutral Glycolipid and Ganglioside Fraction

### 3.1.1. Extraction

The procedure used is essentially that originally described in ref. *4.*

1. To 0.5–1.0 mL of packed cells or up to 1 g wet wt of tissue, add 19.5 mL chloroform:methanol (C:M), 2:1 (v/v), and make a fine dispersion rapidly with the homogenizer.
2. Transfer the suspension to glass centrifuge tubes; cap with aluminum foil and leave to extract at room temperature for at least 1 h with occasional agitation.
3. Centrifuge the tubes at 3000 rpm for 20 min and remove the supernatant carefully (the tissue residue will not be tightly packed), and filter through a glass fiber filter on a scinter under gentle suction.
4. Re-extract the residue, plus any material that may have been transferred to the filter, by dispersion in 10 mL C:M, 1:1 (v/v), cover, and leave at room temperature overnight. After centrifugation filter the supernatant again and combine with the first filtrate.
5. Evaporate the combined extracts to dryness on a rotary evaporator and dissolve the residue in 20 mL C:M, 2:1 (v/v). Remove any insoluble material by filtration (*see* Note 1).
6. Mix the extract gently, avoiding making an emulsion, with 0.2 vol, i.e., 4 mL 0.1*M* KCl, and leave overnight for the two phases to separate.
7. If the separation is not complete, centrifuge the tube gently for 10 min. Then transfer the upper phase carefully to another tube using a Pasteur pipet, taking care not to remove any insoluble interfacial material.
8. Repartition the lower phase at least twice more by mixing with 0.2 vol of TUP.
9. Keep the combined upper phases for the isolation of the ganglioside fraction (*see* Section 3.1.3.), evaporate the lower phase to dryness in a small vessel on a rotary evaporator, and then dry further in a desiccator over phosphorus pentoxide (*see* Section 3.1.2.).

### 3.1.2. Isolation of Neutral and Simple Charged Glycolipids from the Lower Phase Lipid Fraction

1. Completely acetylate the glycolipids by dissolving the dry residue in 0.3 mL dry pyridine followed by 0.2 mL acetic anhydride. Stopper the vessel and leave at room temperature for 18 h.
2. Transfer the mixture with a large volume of dry toluene, at least 50 mL, and co-evaporate on a rotary evaporator.
3. Dissolve the residue in hexane:dichloroethane, 1:4 (v/v) and run through a small bed prepared from 5 g of florisil, packed and washed in the same solvent.
4. Elute the column sequentially with (a) 15 mL hexane:dichloroethane, 1:4; (b) 15 mL dichloroethane; (c) 15 mL dichloroethane:acetone, 1:1; (d) 10 mL dichloroethane:methanol, 1:9; and (e) 15 mL dichloroethane:methanol: water, 2:8:1 (v/v) (*see* Note 2).
5. Evaporate the combined glycolipid fractions and transfer the residue in 0.1–0.2 mL C:M, 2:1 (v/v) to a small tube.
6. Deacetylate the lipids by the addition of 25–50 µL 0.5% sodium methoxide. After 30 min remove excess methoxide by the addition of 0.2 mL ethyl acetate.
7. Add 1–2 mL of water; emulsify the mixture and transfer to a dialysis tube; dialyze overnight at 4°C against a large volume of water. Freeze-dry the contents of the sac in a tared tube for dry weight estimation.

### 3.1.3. Isolation of Gangliosides and Complex Neutral Lipids from the Upper Phase

The gangliosides and some of the complex neutral glycolipids are conveniently isolated from the upper phase using a C18 reversed-phase cartridge. All solvent mixtures v/v.

1. The 3-mL cartridge supplied by Anachem is suitably packed in a syringe that can be fitted through a Luer adaptor to a filter tube so that gentle suction can be applied. Before use, wash the cartridge sequentially with 10 mL methanol and 20 mL C:M (2:1) three times. Wash finally with 10 mL methanol and then equilibrate with 10 mL TUP.
2. Pass the combined, filtered upper phase from the initial extract through the cartridge at least three times, with gentle suction if necessary. Wash the inorganic salts from the column with 10 mL water. Then elute the gangliosides and other lipids from the column with 15 mL methanol and 10 mL C:M (2:1).
3. Evaporate the ganglioside fraction on a rotary evaporator under vacuum, and dissolve the residue in 0.2–0.5 mL C:M (1:1). The cartridge can be

reused up to 10 times if washed with methanol (15 mL) and re-equilibrated to TUP between runs.

### 3.1.4. Isolation of Gangliosides
### and Other Acidic Lipids (see Note 3)

The acidic lipids in the upper phase fraction are conveniently isolated by chromatography on a small column of DEAE-Sephadex.

1. Adjust the solvent composition of the upper-phase lipid fraction to reach C:M:W (30:60:8 [v/v/v]) and a final volume of 5 mL. Pass the solution through a 5-mL bed of DEAE-Sephadex A25, acetate form, and wash the bed with 50 mL of the same solvent.
2. Pool the run-through and wash fractions, which contain the neutral glycolipids from the upper phase.
3. Elute the gangliosides and other acidic lipids from the column with 50 mL C:M:0.8$M$ sodium acetate 30:60:8 (v/v/v).
4. Dry the ganglioside fraction on a rotary evaporator (take care as the fraction approaches dryness as the residue may "spit') and redissolve in a small volume of water. Then either dialyze the fraction or desalt on a small column of G25. A similar procedure can be adopted for the lower-phase lipid fraction, which may contain some simple sulfate esters such as sulfatide.

## 3.2. Estimation of Major Lipid Classes in Fractions

### 3.2.1. Estimation of Total Glycosphingolipid

1. Evaporate aliquots of the diluted standard (0.1–1.0 mL) to dryness under $N_2$ in 10-mL glass stoppered tubes, along with samples containing 50–100 µg of lipid.
2. Dissolve the residues in 0.5 mL dioxane, and add 1 mL 2$N$ $H_2SO_4$; stopper and mix. Then heat at 95°C for 3 h (*see* Note 4).
3. After cooling the tubes in cold water make the hydrolysate alkaline by the addition of 1 mL 24% NaOH.
4. Shake for 1 min with 3 mL ethyl acetate to extract the free sphingosine base. Centrifuge and carefully remove the lower aqueous layer and discard. Wash the ethyl acetate layer twice more by shaking with 2 mL water.
5. Next shake the ethyl acetate extract for 1 min with 2 mL sodium acetate buffer and 0.1 mL of the methyl orange solution, then separate the two phases by gentle centrifugation.
6. Transfer a 2-mL portion of the ethyl acetate phase to a fresh tube and add 3 mL 0.1$M$ $H_2SO_4$; shake the tube again and recentrifuge. Carefully remove the lower pink aqueous layer and read its optical density at 515 nm (*see* Notes 5 and 6).

### 3.2.2. Estimation of N-Acetyl Neuraminic Acid

1. Evaporate samples of the acidic lipid fraction containing up to 0.1 mg of NANA, unknowns and ganglioside standards, in stoppered tubes and add 0.5 mL water. Also prepare solvent blanks and NANA standards, 0.05–0.5 mL (10–100 µg NANA) in a total volume of 0.5 mL (*see* Note 7).
2. Mix each sample with 0.1 mL 0.04$M$ periodic acid. Place the standards, containing free NANA on ice for 20 min and incubate the ganglioside samples, containing bound NANA, at 37°C for 60 min.
3. Place all of the samples on ice; add 1.2 mL of the working resorcinol reagent and mix. Stopper the tubes and after 5 min place them in a boiling water bath for 15 min.
4. Remove the tubes and cool in tap water before adding 1.25 mL tertiary-butyl alcohol. Mix the tube contents to give one phase; warm at 37°C to stabilize the color and measure the absorbance at 620 nm.

### 3.2.3. Estimation of Sulfated Glycolipids

1. Evaporate aliquots of samples and standards containing 0.002–0.035 µmol of sulfated lipid in glass tubes. Include a blank tube containing either no lipid or neutral lipid.
2. Dissolve the residue in 5 mL C:M (1:1) and add 5 mL 25 m$M$ H$_2$SO$_4$ and 1.0 mL Azure A solution.
3. Stopper the tubes and shake well for about 30 s. Centrifuge at a low speed to separate the two phase
4. Remove the lower chloroform phase and read the absorbance at 645 nm. The color yield with sulfatide is about 27.5 OD U/µmol of lipid, and with a range of lipid soluble sulfate esters this value is between 25 and 31 (*see* Note 8).

## 3.3. Analysis of Glycolipids by Thin-Layer Chromatography

### 3.3.1. By HPTLC and Chemical Detection

Basically the procedure for loading samples and for carrying out the separation are as used in all TLC procedures. The text is therefore confined to some methodological recommendations to optimize separation.

1. Most of the detection procedures are quite sensitive, therefore load 1–10 µg of the fraction.
2. Load the samples as small streaks, about 5-mm long, rather than as spots since the $R_f$ for some of the lipids can be quite close to one another.
3. Once loaded, condition the plate to the vapor phase before running for at least 30–60 min (*see* Note 9).

4. To maximize the resolution, particularly of ganglioside mixtures, run the plate twice or even three times. Allow the solvent to run almost to the top of the plate, mark the front and dry the plate in a vacuum desiccator before running again in fresh solvent (*see* Note 10).
5. Detection reagents are of two kinds, general sprays that will detect all lipids and that are generally carbonizing sprays, and those that are specific for carbohydrates:

   Reagent A:   A general carbonizing spray: 0.6 g potassium dichromate in 100 mL 55% (w/v) sulfuric acid (add 58 mL concentrated sulfuric acid to 90 mL water); then heat the plate at 150°C for 20–30 min.

   Reagent B:   A 10% solution of copper sulfate in 8% (v/v) phosphoric acid; then heat at 170°C for 10 min. The plate can be scanned in a densitometer to provide a semiquantitative estimate of the lipids.

   Reagent C:   A nondestructive procedure for lipid detection: 0.01% primulin; the lipid bands can be detected under a UV lamp.

   Reagent D:   A glycolipid-specific spray: 0.2% (w/v) orcinol in $2M$ $H_2SO_4$; then heat at 120°C for 5–10 min.

   Reagent E:   A ganglioside-specific quantitative spray: 10 mL 2% resorcinol, 80 mL conc. HCl and 0.25 mL $0.1M$ copper sulfate made to 100 mL with water. The reagent needs to be made up at least 4 h before use but keeps well at 4°C. After spraying lightly (not enough to wet the plate visibly), cover the plate with a clean glass plate, tape, and place the sandwich, inverted, on a hot plate (about 90–100°C) and leave for 30 min for the color to develop.

### 3.3.2. Detection by Ligand Binding

1. Spot aliquots of the lipid sample under investigation onto either an aluminum or plastic-backed plate (10 × 10 cm), 0.1–0.5 µg as 3–4 mm bands. Further samples at a heavier loading can also be spotted for detection with a colorimetric reagent. Run the plate in the desired solvent and dry in a desiccator (see Note 11).
2. Cut off the reference lanes and sandwich them between clean glass plates. These can be sprayed with the desired detection reagent and heated (*see* Section 3.3.1.).
3. Immerse the remaining portion of the plate in the solution of polyisobutylmethacrylate in a fume cupboard. It takes about 1–2 min to impregnate the plate uniformly. Leave the plate to dry. If required, further divide the plate into smaller strips if it is to be tested against several antibodies or lectins or to be exposed to glycosidases (*see* Note 12).

4. Wet the plate or strips by uniformly spraying with BSA-PBS and immerse in this solution for 1 h to block. Then either immerse or coat with antibody diluted in BSA-PBS and leave at 4°C for 4–16 h, depending on the antibody concentration and affinity. If it is immersed in antibody solution overnight, and placed on a rocker, it should not be in too large a box in order to avoid mechanical damage to the layer.

5. Wash the plate by immersion for 1 min in an excess volume of cold PBS three times, drain and expose to second antibody, at 4°C, for 2–4 h. If the second antibody used is a conjugate then wash the samples as before and develop according to the conjugate used (*see* Note 13–16).

6. If an [125]I-second antibody is used, wash the plate rapidly in three changes of PBS and dry in a desiccator. Process the plates further according to the type of antibody or lectin labeling used. The TLC or strips can be taped to card for ease of handling.

## 4. Notes

1. At this stage it is better to separate the gangliosides and similarly charged lipids from the neutral and simple charged glycolipids by phase partitioning.

2. Fraction (b) contains cholesterol, (c) and (d) the acetylated glycolipids, and fraction (e) the phospholipid.

3. The ganglioside fraction can be further fractionated into the monosialo-, disialo-, trisialo-, and higher sialated lipids on a similar column of DEAE-Sephadex. After the lipid fraction is washed onto the column in C:M:water (30:60:8) [v/v/v] further wash the column with 2 vol of methanol. Elute the gangliosides then with a linear gradient from methanol to 0.4$M$ ammonium acetate in methanol, 800 mL total volume. Alternatively, elute the monosialo fraction with approx 100 mL 0.05$M$ ammonium acetate, the disialo fraction with approx 150 mL 0.08$M$ ammonium acetate, and the trisialo and the higher gangliosides with 0.2$M$ ammonium acetate.

4. The most vulnerable stage in the procedure for the estimation of the sphingosine content is the hydrolysis step. With complex glycosphingolipids with large carbohydrate chains longer hydrolysis times may be required.

5. When reading the optical density, sometimes some ethyl acetate may separate out and cloud the sample. This can be prevented by the addition of a few microliters of ethanol.

6. For lower levels of lipid the volumes can be scaled down to 1/2 to 1/4 of those given, providing a standard curve covering the range from <1 to 7.5 μg (25.0 nmol) of sphingosine.

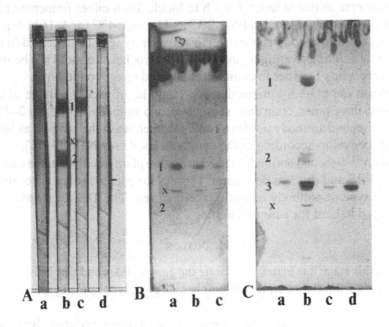

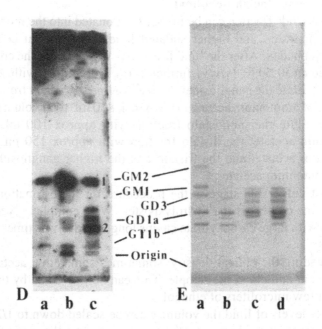

GM2
GM1
GD3
GD1a
GT1b
Origin

7. The procedure for the estimation of NANA is suitable for tissue from the central nervous system where the ganglioside concentration is particularly high, 0.3–1.2 µg NANA/mg wet wt of tissue, however most other tissues contain much less. It may therefore be necessary to use large aliquots of the ganglioside fraction for the estimation or to scale the whole procedure down by reducing the volumes of all reagents by 1/5 or 1/10.

8. For the estimation of the sulfated lipids, if no specific lipid is available as a standard, the values can be expressed in terms of a sodium dodecyl sulfate standard, color yield of 29 absorbance U/µmol.

9. Tanks with a compartmentalized bottom are convenient, the plate being placed in a dry compartment with solvent mixture in the second. After conditioning, fresh running solvent is carefully introduced to the compartment containing the plate.

Fig. 1 . *(opposite page; see* Note 13 for discussion*)* Thin-layer-chromatography, immuno-overlay technique. Mixtures of gangliosides were separated on either plastic-backed TLC sheets (**A** and **B**) or aluminum HPTLC sheets (**C, D,** and **E**) using the solvent system chloroform:methanol:0.5% $CaCl_2$ (60:40:9 [v/v/v]). For primary antibody binding the sera (all human) were diluted 1:100 in 1% BSA-PBS and incubated with the chromatograms overnight at 4°C. Bound antibody was detected by incubating first with class specific rabbit antihuman Ig (1:1000) followed by the stated antirabbit conjugate. (**A**) Human IgM binding to human brain gangliosides, GM1 (1) and GD1b (2) together, serum b and GM1 alone in serum c. Sera a and d are controls. The band x is an artifact from the antirabbit serum. Detection is with peroxidase-conjugated antirabbit developed with 3,3'-diaminobenzidine and $H_2O_2$. (**B**) Anti-GM1/GD1b detected using alkaline phosphatase-conjugated antirabbit, development with 5-bromo, 4-chloro, 3 indolyl phosphate and nitroblue tetrazolium. Lanes a, b, and c loaded with human brain ganglioside 0.5 µg/mm, 0.25 µg/mm, and 0.05 µg/mm, respectively. (**C**) Antibody detection using gold conjugated antirabbit followed by silver enhancement. Human monoclonal IgM binding to sulfatide (1), ($SO_4$3-Galβ1-1Cer); sulfated glucuronyl paragloboside (3), ($SO_4$3-Glucβ1-3Galβ1-4GlcNAcβ1-3Galβ1-4Glcβ1-1Cer), and an unknown lipid (2). Lanes a and b are of human peripheral nerve acidic glycolipids, c and d similar fractions from human fetal brain. (**D**) Bound antibody detection using [125]I-antirabbit and autoradiography. The antibody binds predominantly to ganglioside GM2 (1) and IV[4]GalNAc-GD1a (2) (GalNacβ1-4Gal(3-2α NeuAc)β1-3GalNAcβ1-4Gal(3-2αNeuAc)β1-4Glcβ1-1Cer). Lane a, total human brain ganglioside; lane b, standards GM2, GM1 ,GD1a, and GT1b; lane c, ox spinal cord gangliosides (overloaded). (**E**) TLC of ganglioside mixtures, detected by resorcinol spray. Lane a, standard mixture; lane b, human fetal brain; lanes c and d, 1 and 3 µg ganglioside/mm, respectively. Abbreviations: Gal, galactose; Cer, ceramide; Glc, glucose; GlcNAc, *N*-acetylglucosamine; GalNAc, *N*-acetylgalactosamine; NeuAc, *N*-acetylneuraminic acid; Gluc, glucuronic acid.

10. Sometimes spots appear as close doublets, this generally represents differences in the ceramide portion of the lipid, the carbohydrate being the same. If the neutral lipids are difficult to resolve, it may be an advantage to run them on HPTLC as the acetate derivatives *(5)* prepared as given in Section 3.1.2.

11. The most common problem with the overlay procedure is that the layer may lift off, detection reagents may then be trapped and give spurious spots and bands. The plastic backed plates are intrinsically more stable and can be used even without impregnation with poly-isobutyl-methacrylate if the reagents are applied by overlaying.

12. If there is some difficulty in resolving the lipids then they can be chromatographed as the acetate esters. The mixture is acetylated as given in Section 3.1.2., and the esters hydrolyzed before the application of antibody by exposing the dry plate to concentrated ammonia vapor for 4 h at room temperature.

13. Binding of antibody or lectin can be detected by most standard methods, i.e., peroxidase or alkaline phosphatase conjugates and gold conjugates followed by silver enhancement (Fig. 1). Occasionally at the high pH used in developing alkaline phosphatase the layer may become detached.

14. A large number of antisera and monoclonal antibodies with different carbohydrate specificities are available. Lists of mouse and rat monoclonals *(6)* and human monoclonals *(7)* are given with their haptenic specificities. BioCarb (Lund) specializes in anticarbohydrate monoclonals, and numerous companies sell lectin conjugates and antilectin antibodies.

15. In the examples shown in Fig. 1 the primary antibodies are human, and the second a rabbit class specific antihuman Ig antibody, which after washing as above, is then detected by incubating the chromatogram for a third time in $^{125}$I-antirabbit-IgG, $10^6$dpm/cm$^2$.

16. Further details of the carbohydrate structure can be obtained by the application of specific glycosidases to the plate before using an antibody or lectin *(8)*. This kind of structural approach can be very useful, since "neoglycolipids" can be produced synthetically from glycoproteins and these methods are then applicable to examining even protein carbohydrate structures *(9)*.

## References

1. Lauter, C. J. and Trams, E. G. (1962) A spectrophotometric determination of sphingosine. *J. Lipid Res.* **3,** 136–138.
2. Jourdian, G. W., Dean, L., and Roseman, S. (1971) The sialic acids; IX a periodate-resorcinol method for the quantitative estimation of free sialic acids and their glycosides. *J. Biol. Chem.* **246,** 430–435.

3. Kean, E. L. (1968) Rapid, sensitive spectrophotometric method for quantitative determination of sulphatides. *J. Lipid Res.* **9,** 319–327.

4. Saito, T. and Hakomori, S. (1971) Quantitative isolation of total glycosphingolipids from animal cells. *J. Lipid Res.* **12,** 257–259.

5. Kannagi, R., Watanabe K., and Hakomori S. (1987) Isolation and purification of glycosphingolipids by high-performance liquid chromatography, in *Methods in Enzymology*, vol. 138 (Ginsburg, V., ed.), Academic, New York, pp. 3–12.

6. Magnani, J. L. (1987) Mouse and rat monoclonal antibodies directed against carbohydrates, in *Methods in Enzymology*, vol. 138 (Ginsburg, V., ed.), Academic, New York, pp. 484–491.

7. Spitalnik, S. L. (1987) Human monoclonal antibodies directed against carbohydrates, in *Methods in Enzymology*, vol. 138 (Ginsburg, V., ed.), Academic, New York, pp. 492–503.

8. Saito, M., Kasai, N., and Yu, R. K. (1985) In situ determination of basic carbohydrate structures of gangliosides on thin-layer plates. *Anal. Biochem.* **148,** 54–58.

9. Childs, R. A., Drickamer, K., Kawasaki, T., Thiel, S., Mizuochi, T., and Fiezi, T. (1989) Neoglycolipids as probes of oligosaccharide recognition by recombinant and natural mannose-binding proteins of the rat and man. *Biochem. J.* **262,** 131–138.

3. Kean, E. L. (1968) Rapid sensitive spectrophotometric method for quantitative determination of sulphatides. J. Lipid Res. 9, 319–327.
4. Saito, T. and Hakomori, S. (1971) Quantitative isolation of total glycosphingolipids from animal cells. J. Lipid Res. 12, 257–259.
5. Kannagi, R., Watanabe, K., and Hakomori S. (1987) Isolation and purification of glycosphingolipids by high-performance liquid chromatography. In Methods in Enzymology, vol. 138 (Ginsburg, V., ed.), Academic, New York, pp. 3–12.
6. Magnani, J. L. (1987) Mouse and rat monoclonal antibodies directed against glycolipids. In Methods in Enzymology, vol. 138 (Ginsburg, V., ed.), Academic, New York, pp. 484–491.
7. Spitalnik, S. L. (1987) Human monoclonal antibodies directed against carbohydrates. In Methods in Enzymology, vol. 138 (Ginsburg, V., ed.), Academic, New York, pp. 492–504.
8. Saito, M., Yasui, K., and Yu, R. K. (1985) ... and determination of basic carbohydrate structures of ganglioside on thin-layer plates. Anal. Biochem. 148, 54–58.
9. Childs, R. A., Dalchau, R., Kruckeberg, T., Thorpe, T., and Feizi, T. (1985) ... monoclonal is an example of glycoprotein recognition by recombi...

3. Kean, E. L. (1968) Rapid, sensitive spectrophotometric method for quantitative determination of sulphatides. *J. Lipid Res.* **9,** 319–327.

4. Saito, T. and Hakomori, S. (1971) Quantitative isolation of total glycosphingolipids from animal cells. *J. Lipid Res.* **12,** 257–259.

5. Kannagi, R., Watanabe K., and Hakomori S. (1987) Isolation and purification of glycosphingolipids by high-performance liquid chromatography, in *Methods in Enzymology*, vol. 138 (Ginsburg, V., ed.), Academic, New York, pp. 3–12.

6. Magnani, J. L. (1987) Mouse and rat monoclonal antibodies directed against carbohydrates, in *Methods in Enzymology*, vol. 138 (Ginsburg, V., ed.), Academic, New York, pp. 484–491.

7. Spitalnik, S. L. (1987) Human monoclonal antibodies directed against carbohydrates, in *Methods in Enzymology*, vol. 138 (Ginsburg, V., ed.), Academic, New York, pp. 492–503.

8. Saito, M., Kasai, N., and Yu, R. K. (1985) In situ determination of basic carbohydrate structures of gangliosides on thin-layer plates. *Anal. Biochem.* **148,** 54–58.

9. Childs, R. A., Drickamer, K., Kawasaki, T., Thiel, S., Mizuochi, T., and Fiezi, T. (1989) Neoglycolipids as probes of oligosaccharide recognition by recombinant and natural mannose-binding proteins of the rat and man. *Biochem. J.* **262,** 131–138.

3. Kean, E. L. (1968) Rapid, sensitive spectrophotometric method for quantitative determination of sulphatides. J. Lipid Res. 9, 319–327.

4. Saito, T. and Hakomori, S. (1971) Quantitative isolation of total glycosphingolipids from animal cells. J. Lipid Res. 12, 257–259.

5. Kannagi, R., Watanabe, K., and Hakomori, S. (1987) Isolation and purification of glycosphingolipids by high-performance liquid chromatography. In Methods in Enzymology, vol. 138 (Ginsburg, V., ed.), Academic, New York, pp. 3–12.

6. Magnani, J. L. (1985) Mouse and rat monoclonal antibodies directed against glycolipids. In Methods in Enzymology, vol. 138 (Ginsburg, V., ed.), Academic, New York, pp. 484–491.

7. Spitalnik, S. L. (1987) Human monoclonal antibodies directed against carbohydrates. In Methods in Enzymology, vol. 138 (Ginsburg, V., ed.), Academic, New York, pp. 492–504.

8. Saito, Nakasaki, H., and Yu, R. K. (1985) In situ determination of basic carbohydrate structures of gangliosides on thin-layer plates. Anal. Biochem. 148, 54–58.

9. Childs, R. A., Dalchau, R., Kawasaki, T., Ebert, T., Mizuochi, T., and Feizi, T. (1995) Multivalency is essential for the recognition of the tetrasaccharide repeating...

# Appendix

## 1. Density Gradient Media

Table 1
Concentration, Density, and Refractive Index of Sucrose Solutions

| Concentration | | | | |
|---|---|---|---|---|
| %, w/w | %, w/v | Molarity | Density, $g/cm^3$ | Refractive index |
| 0 | 0 | 0 | 0.998 | 1.3330 |
| 5.0 | 5.1 | 0.15 | 1.018 | 1.3403 |
| 10.0 | 10.4 | 0.30 | 1.038 | 1.3479 |
| 15.0 | 15.9 | 0.46 | 1.059 | 1.3557 |
| 20.0 | 21.6 | 0.63 | 1.081 | 1.3639 |
| 25.0 | 27.6 | 0.80 | 1.104 | 1.3723 |
| 30.0 | 33.8 | 0.99 | 1.127 | 1.3811 |
| 35.0 | 40.3 | 1.18 | 1.151 | 1.3902 |
| 40.0 | 47.1 | 1.38 | 1.176 | 1.3997 |
| 45.0 | 54.1 | 1.58 | 1.203 | 1.4096 |
| 50.0 | 61.5 | 1.80 | 1.230 | 1.4200 |
| 55.0 | 69.2 | 2.02 | 1.258 | 1.5307 |
| 60.0 | 77.2 | 2.25 | 1.286 | 1.4418 |
| 65.0 | 85.5 | 2.50 | 1.316 | 1.4532 |

Data from Griffith, O. M. (1986) *Techniques of Preparative, Zonal, and Continuous Flow Ultracentrifugation*, 5th ed. Beckman Instruments Inc., Spinco Division, Palo Alto, CA

From: *Methods in Molecular Biology, vol. 19: Biomembrane Protocols: I. Isolation and Analysis*
Edited by: J. M. Graham and J. A. Higgins Copyright ©1993 Humana Press Inc., Totowa, NJ

Table 2
Concentration, Density, and Refractive
Index of Nycodenz™ Solutions

| Concentration | | Density, g/cm$^3$ | Refractive index |
|---|---|---|---|
| %, w/v | Molarity | | |
| 0 | 0 | 0.998 | 1.3330 |
| 5 | 0.061 | 1.025 | 1.3412 |
| 10 | 0.122 | 1.052 | 1.3494 |
| 15 | 0.183 | 1.079 | 1.3577 |
| 20 | 0.244 | 1.105 | 1.3659 |
| 25 | 0.304 | 1.132 | 1.3742 |
| 30 | 0.365 | 1.159 | 1.3824 |
| 35 | 0.426 | 1.186 | 1.3906 |
| 40 | 0.487 | 1.212 | 1.3988 |
| 45 | 0.548 | 1.238 | 1.4071 |
| 50 | 0.609 | 1.265 | 1.4153 |
| 55 | 0.670 | 1.292 | 1.4326 |
| 60 | 0.731 | 1.319 | 1.4318 |
| 65 | 0.792 | 1.346 | 1.4400 |
| 70 | 0.853 | 1.372 | 1.4482 |
| 75 | 0.914 | 1.399 | 1.4565 |
| 80 | 0.974 | 1.426 | 1.4647 |

Data from Rickwood, D. (1983) Properties of iodinated density gradient media, in *Iodinated Density Gradient Media: A Practical Approach* (Rickwood, D., ed.), IRL Press at Oxford University Press, Oxford, UK, pp. 1–21.

## 2. Balanced Salt Solutions

Dulbecco's Phosphate Buffered Saline (PBS): 8.00 g NaCl; 0.20 g KCl; 0.10 g $CaCl_2$; 0.10g $MgCl_2 \cdot 6H_2O$; 1.15 g $Na_2HPO_4 \cdot 2H_2O$; 0.20 g $KH_2PO_4$ made up to 1 L with distilled water. Sometimes the $Ca^{2+}$ and $Mg^{2+}$ salts are omitted.

Hank's Balanced Salt Solution (HBSS): 8.00 g NaCl; 0.40 g KCl; 0.14 g $CaCl_2$; 0.10 g $MgSO_4 \cdot 7H_2O$; 0.10 g $MgCl_2 \cdot 6H_2O$; 0.06 g $Na_2HPO_4 \cdot 2H_2O$; 0.06 g $KH_2PO_4$; 1.0 glucose; 0.35 g $NaHCO_3$; 0.02 g phenol red; made up to 1 L with distilled water.

A number of saline solutions, in which the phosphate buffer is replaced by an organic one, are widely used, but their precise composition varies from laboratory to laboratory. Tris tends to be avoided, since it appears to be toxic to some types of mammalian cell; HEPES and TRICINE are popular alternatives, used at 5–10 m*M* concentration.

# Index